The A
of
Koi Keeping

The Art of Koi Keeping

A COMPLETE GUIDE

Peter Cole

BLANDFORD

A BLANDFORD BOOK
First published in paperback 1993 by Blandford
An imprint of Cassell
Villiers House, 41-47 Strand
London WC2N 5JE

Original hardback edition 1990

Distributed in the United States by
Sterling Publishing Co. Inc.
387 Park Avenue South, New York, NY 10016-8810

Distributed in Australia by
Capricorn Link (Australia) Pty Ltd
PO Box 665, Lane Cove, NSW 2066

British Library Cataloguing in Publication Data

Cole, Peter
The art of Koi keeping,
1. Pets. Koi, Care
I. Title
639.3′752

ISBN 0-7137-2141-3 (hardback)
0-7137-2387-4 (paperback)

Typeset by Nene Phototypesetters Limited, Northampton
Printed and bound in Hong Kong by Colorcraft Ltd

Dedicated to Ann

Contents

Acknowledgements

I should like to thank the following people for their help with my book: Allan and Jan Chandler of Southern Koi Supplies; Ron ('the pond') Phillips of Aquatic Landscapes; Dr Pierre de Kinkelin (Ministry of Agriculture) France; Galerie Michel Mathonnet, Paris; Corinne and Adeline of *Canicule*. Also, special thanks to André and Syn Zagradsky.

The following people have also supplied photographs: A. Chandler, R. Phillips, Sky Photographic Services and members of the British Koi Keepers' Society, together with Dr Pierre de Kinkelin, M. Dorson and S. Chilmonczyk, who supplied microphotography.

The sketches and line drawings were done by A. Zagradsky.

Above and opposite: In the beginning, carp that possessed scales of a different colour appeared naturally, and it is from these mutant fish that the Japanese have produced the colourful examples of Koi encountered today.

Preface

This book has been written for those of you who have already discovered these beautiful fish from Japan – and for the many more who will come to know them as a result.

I hope sincerely that it will enable you to understand the art of Koi keeping, and that your participation will be rewarded with many years of the happiness and pleasure that this hobby brings.

Peter Cole

History

Records of Koi culture go back more than 1000 years and the fish are known to have first been kept by noblemen and Samurai. However, the artistic development that has established the majority of today's colourful examples is said to have taken place at the turn of this century. Some of the colours and patterns have taken many Koi generations to perfect. In Japan, Koi are regarded as a symbol of masculinity and love.

The carp makes a significant appearance in Japanese art and legends. It represents tenacity and is depicted as tackling courageously the strongest currents and jumping the highest cascades. Often, it is seen in the form of ornaments and is displayed in Japanese homes during the festival of T*an-go-no-sek-ku*, which announces and celebrates the coming of age of the young man of the household.

Opposite: Nishikigoi, or Koi, are descendants of the common carp, *Cyprinus carpio*, which are cultivated for food in Japan. They are not, as some believe, coloured goldfish.

Beginning

To introduce Koi into any pond without taking into consideration the condition of the water is not fair on the Koi, and may well lead to great disappointment if in consequence they fall sick and even die.

Koi are much larger and grow more rapidly than any other freshwater ornamental fish, using more oxygen and producing greater quantities of waste products. This waste is basically ammonia and carbon dioxide produced by, respectively, digestion and respiration. It has to be kept below a critical level that can eventually harm your Koi. This control is achieved by constant physical and biological filtration of the pond water, and by occasional removal of any debris that has sunk to the bottom of the pond.

A little knowledge and thought when setting up your pond will eliminate most of the problems that can be encountered in Koi keeping, and help to ensure many years of trouble-free pleasure.

A garden pond with its characteristic green water.

Koi are hardy fish if kept under ideal conditions.

Pond and Filter Construction in General

One of the essentials of Koi keeping is to establish and maintain suitable water conditions. Apart from the temperature difference involved in keeping exotic aquarium fish, the requirements are much the same: an efficient filter system, a reliable oxygen supply, and a means of keeping the bottom of the pond clean. These are all indispensable for maintaining ideal water conditions.

To remain in good health, your Koi will depend on your care and their environment. Therefore thoughtful planning before you start to construct your pond is very important. In return, your Koi will bring you many years of peaceful pleasure for very little effort.

When choosing a site for your pond, take into consideration the following important factors:

1 A pond sited near the house will prove more entertaining for your family and visitors than one located at the far end of the garden.

2 Overhanging trees will shed their leaves and pollen into your pond, fouling the water and possibly harming your Koi's health. If necessary, trees can be cut back and the edges of the pond can be constructed so as to incorporate a cover that can be put into place before the autumn. However, do not sacrifice trees for your pond!

A surface skimmer can be installed during construction, provided that an external filter system is to be used or the skimmer has its own pump.

3 The pond will require only a moderate amount of sunlight. Excessive exposure to the sun will not only turn the water green but may lead to sunburn, causing your Koi's colours to fade or damaging their skin. A shaded area can be created by using plants.
4 Ideally, the pond should be sheltered from easterly winds to help stabilize the water temperature. Shelter can be provided by a nearby wall or house. If necessary, the earth removed during the pond's construction can create shelter as well as a raised ornamental garden. If you are planning a waterfall, this too should be protected from easterly winds to avoid cooling the water.

Before deciding on the size and shape of your pond, consider the following points:
1 The size and shape of your pond will influence the types of materials required and the total cost.
2 Do not make the pond too small, since then you will want to enlarge it later. It will not be expensive to increase its area during the initial construction. However, to enlarge a pond that has already been installed is not only difficult, but can cost as much as building a new pond.
3 Do you want a pond with a formal or an informal shape?
4 Make your pond deep enough to provide ample water and good wintering facilities for your Koi. A shallow pond will subject your Koi to rapid, stressful temperature changes. A deeper pond will keep the water temperature stable and warm enough for your fish to settle on the bottom during cold weather. The pond's depth will also have an influence on the intensity of the Koi's colours and their growth rate. You must allow a difference in height of about 25cm (10in) between the edge of the pond and the proposed water level, to prevent your Koi from jumping out. If the edge of the pond is raised by this amount, it will also prevent dust and dirt from falling into the water. The minimum water depth should ideally be 1–1.25m (3–4ft).
5 An overflow and possibly some means of emptying the pond should not be overlooked during construction. A shallow area for feeding should be incorporated, preferably at the approach to the pond. This is particularly recommended if there is any likelihood of children falling in. Where small children are at risk,

Some Koi keepers build their gardens around their ponds.

it is further advisable to build a sturdy fence around the perimeter or even to construct instead a raised formal pond.

6 Which type of filter system will you use? The internal type are inexpensive to build and easy to install. They can also create a shallow area that will take up part of the pond's overall water volume rather than extra space outside. The efficiency of external filter systems can be improved by incorporating such refinements as surface skimmers and bottom-drain settling chambers. However, this type of system is more expensive to build and takes up space outside of the pond. Ready-made filter modules can be obtained from a Koi dealer or garden centre. If your pond is large enough, several modules can be linked together to create an efficient system.

Whichever system you choose, it must be powerful enough to cope with the initial volume of water and the eventual total number of Koi you plan to keep, bearing in mind that adult Koi can reach 60–70 cm (24–28 in) in length. (Water volume × 3 Koi = 1 cubic metre.)

7 The bottom of the pond will have to be cleaned regularly.

A formal-shaped pond can easily become part of an existing patio or terrace, and can be raised above ground-level. An informal-shaped pond can be landscaped into a garden by using the earth removed during its construction.

Will you install a bottom drain or will you use a vacuum pump? A bottom drain will enable you to clean the pond easily, by periodically operating a gate valve or stand-pipe to disperse the sunken debris into a drainage system. The water removed in this way will have to be replaced. Remember, when you refill the pond, the water should be directed onto the surface with a spray. This will vaporize any chemicals – chlorine, for example – that have been used to purify the water.

A bottom drain can be connected directly to a settling chamber incorporated in an external filter system. This set-up provides continuous cleaning action and only the settling chamber will have to be cleaned periodically. It combines efficiency with a minimum water loss. In some cases, the bottom of the pond can be cleaned more thoroughly with a vacuum pump. Whichever method you choose, it is imperative that you clean the bottom regularly, since failure to do so will threaten the health of your Koi.

8 Which type of pump are you going to use to circulate the pond water through the filter system? In an average pond, it is necessary to circulate approximately a quarter of the total water volume through the filter system each hour.

There are two types of pump: the submersible type or the surface type. Submersible pumps are easier to install and versatile, but consume a lot of electricity. Surface central-heating circulating pumps are more popular. They take a little longer to install, but consume very little electricity. Most surface models have adjustable outputs and are very reliable.

Bear in mind that the pump will be in permanent use. Make sure that the model you choose has power in reserve, since reducing the output is easy, but buying a bigger pump is expensive.

This pump circulates about 10,000 litres of pond water. It has three output settings, the lowest being used in winter and the intermediate throughout the summer. The highest setting is used to drive the water up and over a waterfall.

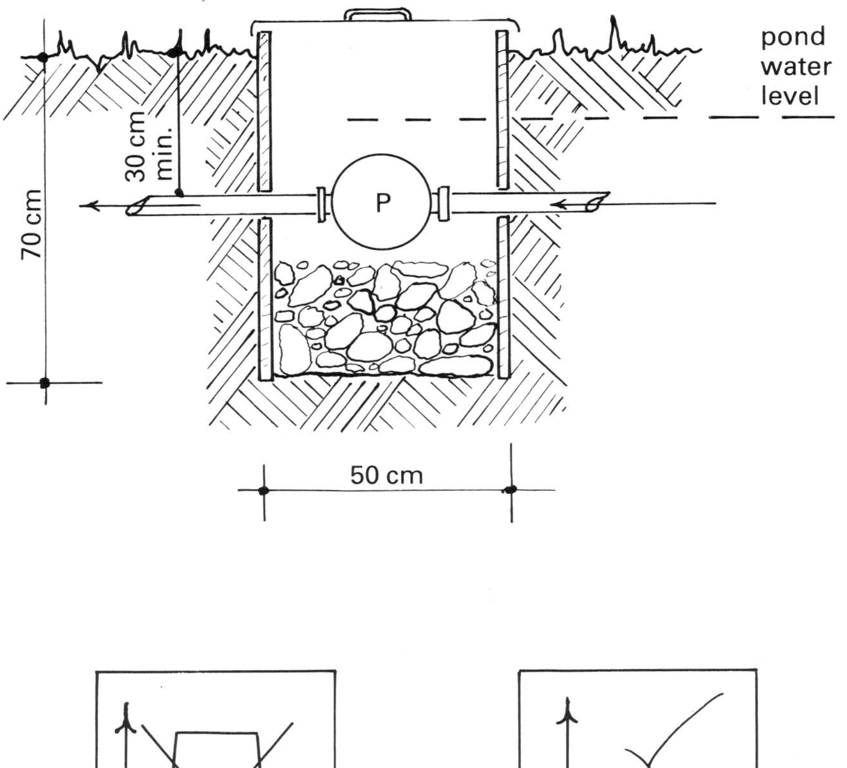

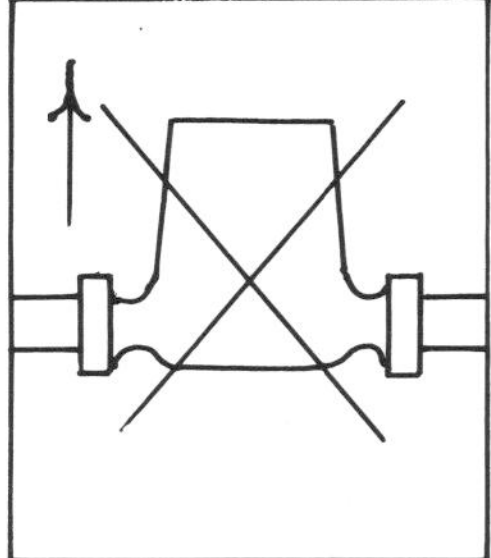

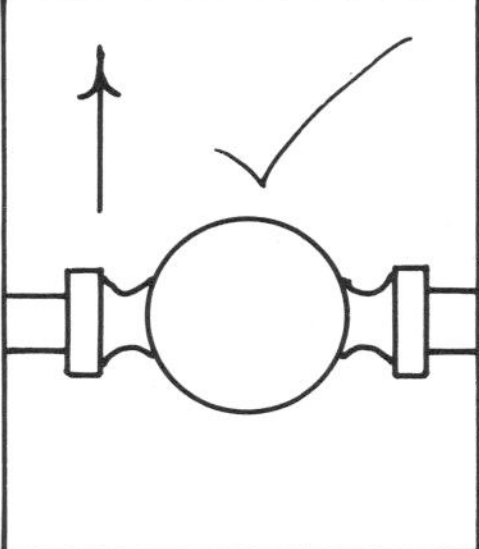

Fig 1 The pump body should always be mounted in a horizontal position and bled regularly to remove air.

Ponds

Pond construction can be carried out at any time of the year, although the best time is the summer or early autumn, with more daylight hours and the ground not too heavy with moisture – digging being the hardest part. If you are planning to construct a large pond, it might be advisable to seek help or hire a mechanical digger to save time and possibly your back.

Several types of material can be used to build your pond and each has its advantages. The most popular materials currently used by amateurs and professionals alike are PVC and butyl rubber. Many agricultural reservoirs are created with these materials, which are simple to use, do not dry or crack, and are resistant to ultra-violet rays and sub-zero temperatures. Leaks are uncommon and easy to repair. Most butyl liners carry a maker's warranty of up to 15 years, although their life expectancy is said to be more than 25 years.

Liners in sizes larger than standard can be ordered from the supplier or manufacturer. Although joining tape is available that is effective for joins or repairs on a clean, dry surface, large joins or special shapes should be made by the manufacturer or supplier, using a heat-welding method.

When calculating the liner's size, allow an approximate 30 cm (12 in) overlap for securing its edges around the perimeter of the pond. The size should be calculated as follows:
Liner length = depth × 2 + length + 2 × overlap.
Liner width = depth × 2 + width + 2 × overlap.

Ponds can be built using PVC or butyl-rubber liners, or with concrete or fibreglass.

Here are a few tips on using a PVC or butyl-rubber liner:

1 With firm soil, the walls of the pond can rise at 90° from the bottom. With unfavourable soil, the angle should not exceed 60°. If necessary, support the walls with breeze blocks.

When digging out the pond, allow an extra 10–15 cm (4–6 in) in depth for a layer of clean, soft sand to protect the liner from sharp stones. This precaution also applies to the bed of an internal filter system, if one is to be fitted.

2 If a bottom drain is to be installed, the bottom of the pond should slope gently towards the area where the drain will be situated. The drain should be secured in a bed of concrete to prevent it moving under the pressure of the water. The openings of the drain should be set horizontally at the same level as the sand.

3 Where necessary, the walls can be lined with a strong underlay or polystyrene sheeting to protect the liner from sharp stones. The underlay should be of a synthetic, rot-proof material.

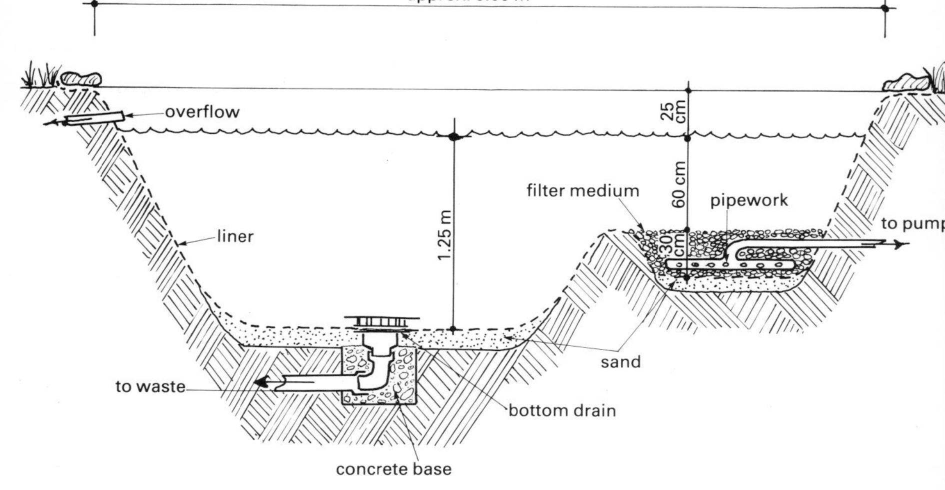

Fig 2 A liner can be used to create a pond that is formal or informal in shape.

Bottom drain

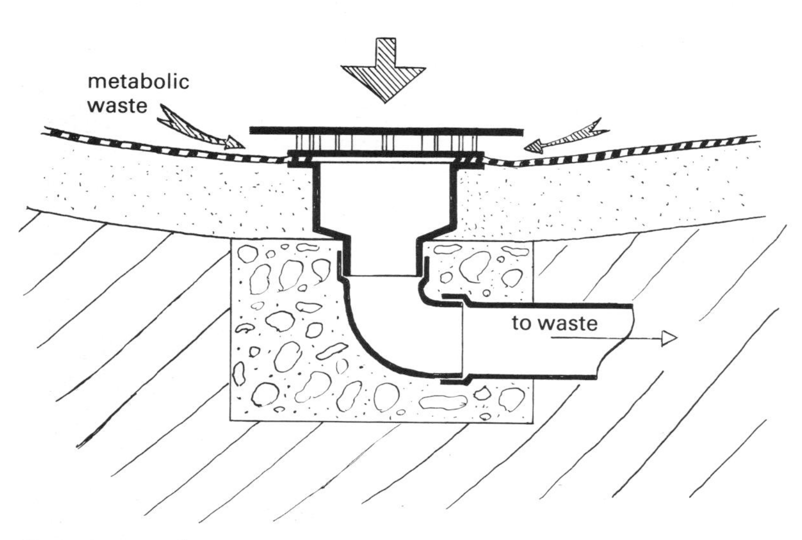

Fig 3 Bottom drain. Water pressure creates cleaning action while operating a gate valve or stand-pipe to waste.

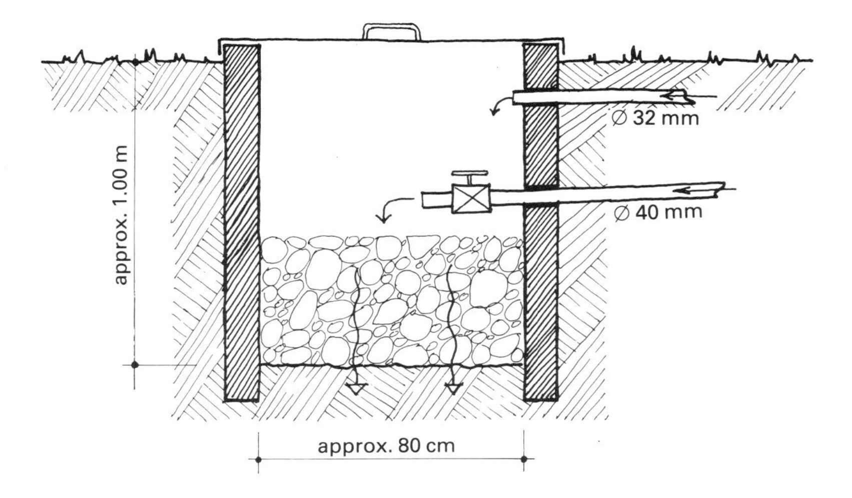

Fig 4 Discharge box for bottom drain and overflow.

A typical bottom drain.

This pond was built using a PVC liner. A bottom drain was connected to the settling chamber of the external filter. A 15 cm (6 in) pipe between the pond and the filter at the surface acts as a surface skimmer.

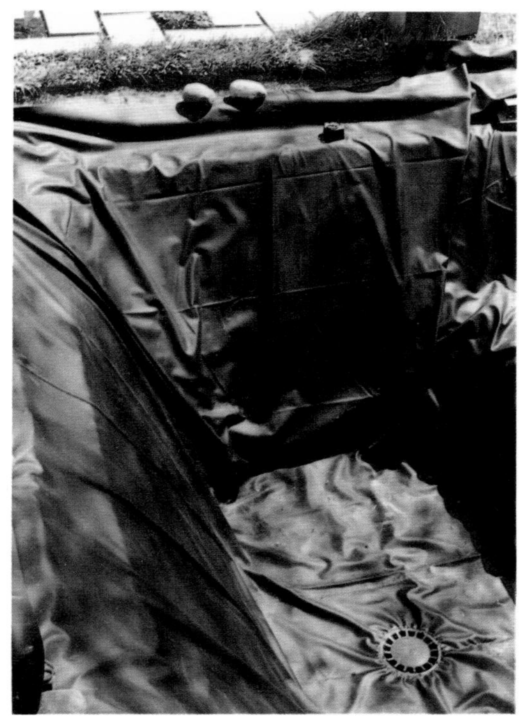

4 To avoid tearing the liner, before unfolding and laying it out all sharp objects should be removed from the surrounding area. The edges of the pond and any irremovable objects can be covered temporarily with a suitable means of protection.

5 After the liner has been centred correctly, the bottom drain, if installed, can be secured. The hole for the drain should be marked out to allow the liner to overlap the inside of the hole by approximately 2cm (¾in). Use a good pair of scissors (never a knife) to cut the hole. The liner can now be lifted and silicone-mastic sealer applied around the mating edge of the drain opening, around the perimeter underneath and on top of the hole in the liner and, most important, on the mating surface of the sealing ring. Press the liner firmly onto the base of the drain, centre the sealing ring and press it into place, and screw down the sealing ring diagonally with even pressure.

To ensure a perfect seal, avoid getting sand on any of the surfaces while using the silicone-mastic sealer and allow at least 24 hours for the sealer to dry. To avoid disturbing the sand underneath the liner while connecting the bottom drain, stand on a flat piece of wood with its sharp corners and rough edges removed.

Note To allow for stretching, do not make at this stage holes for pipe connections or surface skimmers.

Plants can be introduced while the pond is being filled with water.

During the 'start-up' period the pond water may turn green with algae. Eventually, however, beneficial bacteria will become established in the filter bed and the algae will be starved out of existence. At the early stage it is wise to introduce only a few, inexpensive, Koi into your pond.

6 While filling the pond, gently stretch out any creases in the liner and make any necessary folds. Stop filling when the water level reaches approximately 15cm (6in) below the point in the liner where you need to make the next hole. Make your pipe connections with silicone-mastic sealer and allow 24 hours before continuing the installation. Continue to fill the pond and make the necessary pipe connections as described above until the water reaches the required level. Connect the overflow pipe.
7 If you are installing an internal filter system, again to allow for stretching, this needs to be in place while you are filling up the pond. To avoid leaks, clean, soft sand should be placed underneath the liner to protect it from sharp stones. A strong plastic sheet should be placed underneath the filter pipework, before filling the filter bed with medium.
8 Once the pond has been filled with water, the edges of the liner can be secured with heavy stones or slabs. Allow a minimum of 24 hours before introducing any Koi.

This pond was built using a butyl-rubber liner and incorporates an internal filter system. The bottom is cleaned regularly by means of a vacuum pump.

Concrete

One of the advantages of using concrete is that there is almost no restriction on the shape of your pond. As cement contains chemicals that are toxic to fish, a new pond will have to be treated after construction.

Here are a few tips on using concrete:

1 In calculating the depth of the pond, allow for the thickness of the concrete.
2 The sides of the pond should not be at an angle greater than 45°. This precaution, together with a lining of wire mesh, will help to keep the concrete in place while you are creating the walls. The mesh will also help to reinforce the structure.
3 All pipes and fittings should be put into place beforehand. (Do not forget the overflow.)
4 If you prefer, the concrete can be ordered ready mixed. This will save a lot of time and give you a more consistent mixture.

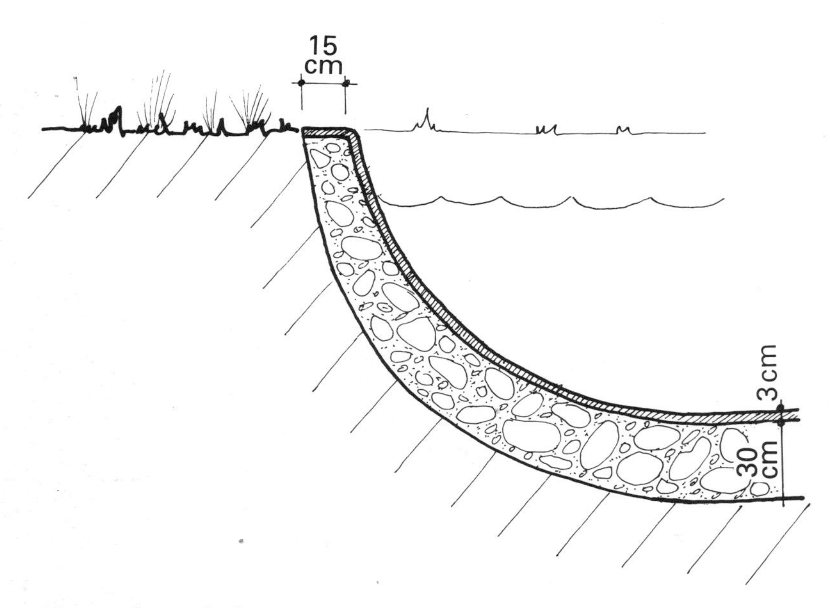

Fig 5 When calculating the amount of concrete you will need, allow no less than 30cm (12in) for the bottom of the pond and walls that should taper to about 15cm (6in) at the top.

5 After about three days, the concrete will be dry enough to enable you to render the interior with a mixture of clean, soft sand and cement to which a waterproofing agent has been added. As a render, apply an even 3 cm (1¼ in) layer of cement. This layer can be made smooth by rubbing gently with a damp sponge or polystyrene block.
6 When the cement is completely dry, the interior of the pond should be treated with a waterproof sealer. This should not be diluted and a second coat should be applied after 48 hours. Black sealer is best for enhancing the colours of the Koi.

Note Use a non-toxic sealer that has been specially adapted for this purpose.

7 Allow a minimum of four days for the sealer to dry and allow two more days after filling the pond with water, before introducing any Koi.

An abandoned concrete pond can be renovated by cleaning it and applying two coats of waterproof sealer. Large cracks can be filled with cement before sealing; minor cracks can be repaired with a 50/50 mixture of clean, soft sand and sealer.

These ponds were built with concrete in 1910 and have been successfully renovated with waterproof sealer. They are sited on sloping ground and are in fact three separate ponds that flow into each other via waterfalls. The middle pond has been divided into four separate chambers to create a rise-and-fall filter system. The lowest, and largest, has had an internal filter system added, which acts as a pre-filter and also removes floating debris and leaves. The top pond holds a variety of goldfish, while the largest pond holds a group of 25 healthy Japanese Koi.

A formal or informal-shaped pond can be constructed by building the walls at a 90° angle in breeze blocks and/or bricks. The interior can be rendered and coated with waterproof sealer. Alternatively, the interior can be lined with polystyrene sheeting and a PVC or butyl-rubber liner. Fibreglass can also be used as a waterproof sealer. Remember that the opposing walls should be horizontal to each other.

The pond illustrated became part of an existing terrace. The double wall provides a decorative finish, together with insulation to help stabilize water temperatures. The slate-built stepped waterfall adds the final touch to the construction and lends 'ambiance' to the surroundings.

Fibreglass is a very robust material and will allow you to create a pond of any shape. The walls should not be steeper than 60°. This will provide an agreeable surface to work on. A thin layer of cement is applied to the walls as a surface to which fibreglass can adhere.

Three layers of fibreglass were applied, the last being lightly sanded and painted with a 'Gel Coat', to give an extra-smooth finish.

Pre-Fabricated Ponds

In the past the hobbyist's choice in a pre-fabricated pond was limited to those shallow fibreglass designs that were very good as children's paddling pools, but completely unsuitable to house fish. However, we are now able to obtain fibreglass and vacuum-formed plastic ponds that are designed with fish and plants in mind. There are models available that cater for the requirements of most hobbyists, provided, of course, that you are happy with the shape. Many of these include adequate depth, a middle shelf to incorporate an internal filter system, if required, and sufficient shelving for a variety of marginal plants.

In proportion to their size, some models can be a little more expensive if compared to the pond-building techniques previously described, but of course as for any of these and the materials used, there is the possibility of less hidden costs if all is well planned and carefully chosen beforehand. Through these aspects alone you could make considerable savings during the installation of any pool.

If you have decided on purchasing a pre-formed pool, here are some important aspects that must be taken into consideration:

1 Size: will you be able to get the pool into the garden without having to take out the front and back door frames, or knock down the neighbours' fence . . . ?
2 If necessary, leave enough space to install an external filter system or, if required, chose a model that has been designed to

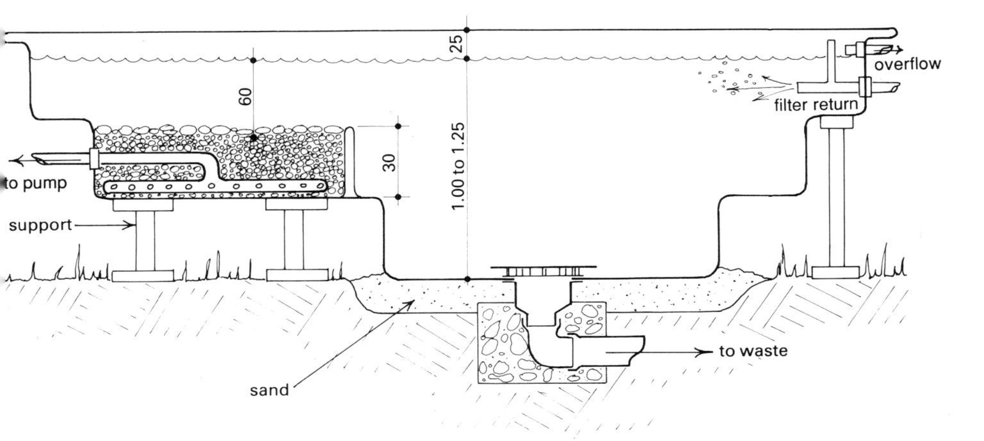

Fig 6 Most fibreglass pools and some plastic designs can be freestanding if you provide support at the necessary strategic points.

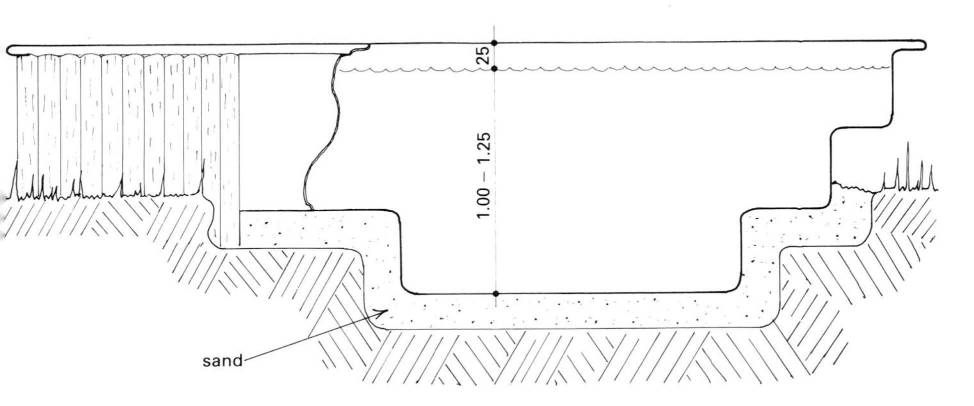

Fig 7 Pre-formed fibreglass pools can be installed at any required depth; most plastic economy versions, however, will need to be encased in earth or sand with a surrounding wall, or partially or totally submerged.

adapt an internal filter. Make careful plans for your pipes and fittings – do not forget the overflow!

3 Any earth from an excavation will have to be removed by skip or used to landscape the pool into the surrounding garden.

4 As for liners, to prevent penetration from underneath, a pre-formed pond will have to be settled on a layer of clean, soft

sand. This will also help you to adjust and make sure that the top edges remain level.
5 Nearly all fibreglass pools and some plastic models can be freestanding with a minimum of support. This method will require some form of insulation to stabilize water temperatures. For example, if logs or bricks and stones are being used to form a wall around the surrounding edges, it would be advisable to fill in with earth or a suitable insulating material.
6 Quality fibreglass pools are available in any colour. Black would be preferable to enhance the colours of the fish.
7 Make sure that the pool you have chosen is deep enough to over-winter your Koi. . .

Once installed, a surrounding wall can be built using logs, bricks or stones.

Filters

Filtering your pond water is, with feeding, the most important part of Koi keeping. Without effective filtering, you may well encounter so many problems that your hobby becomes neither pleasant nor successful. Properly filtered water will not only be clean, but will also keep your Koi in healthy condition.

Filtration is achieved by passing the pond water through a container filled with a filter medium such as gravel, to separate any floating debris that will accumulate and cloud the water. If this action is continuous, bacteria will eventually grow on the surface of the filter medium and break down any dissolved metabolic waste such as ammonia and nitrites into less harmful products by a process known as 'nitrification'. Once this beneficial flora or bacteria is established, the action of the filter then becomes biological, supplying your pond continuously with clean, fresh water.

It will take several weeks for the bacteria to become established, and so you should introduce only a few Koi during this period. There is a possibility that the water will turn green with the growth of algae during the first few weeks of operation of a new system. Eventually, however, the biological action of the filter will make the water clear as it starves the algae out of existence. Do not introduce chemicals to clear the algae, since they could destroy the bacteria growing in the filter bed. Various methods can be used to accelerate the accumulation of bacteria. In the past, rotting prawns or decaying fish food were added to the filter medium, but nowadays instant dried bacteria culture is used.

Filtration should be continuous throughout the year, even during the winter. If for any reason the flow of water through the filter is stopped, within two to three days the beneficial bacteria will disappear, through lack of oxygen, and be replaced by bacteria that will prove harmful to your Koi. If this happens, the filter will have to be cleaned out before restarting the circulation of water, and a new population of bacteria established before filtration will begin again.

Note Some fish medicines can destroy the bacteria in the filter.

Koi keepers who are frequently absent should consider a back-up water circulating system and arrange constant surveillance by a trustworthy friend or neighbour.

Various materials can be used as an effective filter medium. The basic requirements are a rough surface to which the bacteria can cling, and particles not so small as to become clogged. Ideally, these should measure approximately 15–20 mm (5/8–3/4 in). The popular mediums are gravel with a rough surface and lightweight aggregates.

Lightweight aggregates are ideal in an internal or external filter system in cases where a pond liner has been used. A pre-filter or settling chamber incorporated in an external filter system can use filter brushes to strain off any particularly large floating debris.

Bacteria colonies spread out horizontally, since they require surface area rather than depth. Thus, 30 cm (12 in) of filter medium spread over an area of 30–45 per cent of that of the pond itself, will establish a large enough bacteria colony to produce effective filtration. This will only happen, however, if the stocking rate is not too high and if a minimum of approximately one quarter of the total volume of pond water passes through the filter medium each hour. These figures are based on a stocking rate of three adult Koi per cubic metre (220 gallons imperial/264 gallons US) of pond water. Higher stocking levels would require a greater volume of water and/or a higher flow rate and filter area.

A basic requirement for efficient filtering is a good oxygen supply both before and after the water passes through the filter system. During the nitrification process, approximately 90 per

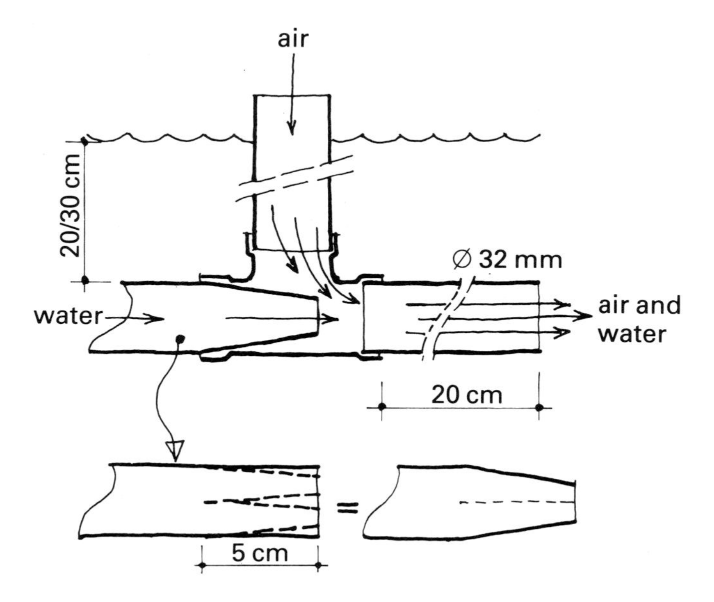

Fig 8 A venturi can be made from PVC piping and fittings.

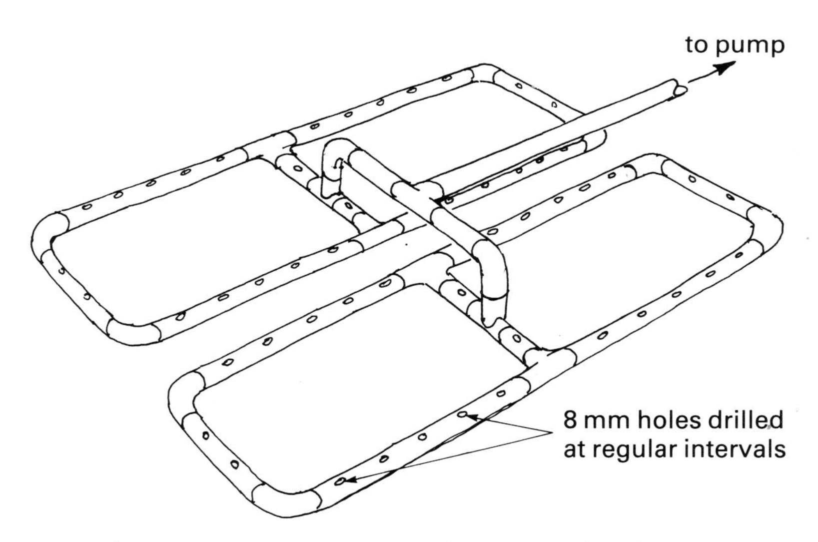

Fig 9 The piping network is connected to a pump that draws the pond water through the filter medium to create filter action.

Your Koi's health, colour and growth rate will all benefit greatly from an efficient filtration system.

A venturi will circulate the pond water and in doing so direct the sunken debris into one area. The gentle current will provide exercise for the Koi and also keep an area free from ice should the surface freeze in winter.

This excellent filter system was built in part with breeze blocks. The baffle boards and perforated plates were made from marine ply.

cent of the dissolved oxygen is used by the beneficial bacteria, so that the oxygen must be replaced before the water re-enters the pond. This can be achieved by passing the water over a waterfall, or through a fountain or a venturi. The latter is the most effective method. A waterfall or fountain should be considered as an 'extra', since each needs to be turned off during the evenings and winter months to avoid unnecessary cooling of the water.

A variety of materials can be used to build your own filter system. An internal filter system can be made from a network of drilled plastic piping that is covered with a filter medium. Marine ply lined with PVC or fibreglass, and prefabricated plastic or fibreglass tanks, can all be transformed into effective filter systems. Breeze blocks rendered with waterproof cement and coated with waterproof sealer can provide a series of filter chambers for a highly efficient filter system. Baffle boards or perforated support plates for filter medium can be made from marine ply, rigid plastic or fibreglass sheeting.

An added refinement to pond filter systems are the sand filters used for swimming pools. Although they provide additional water clarity, they are likely to become clogged if not

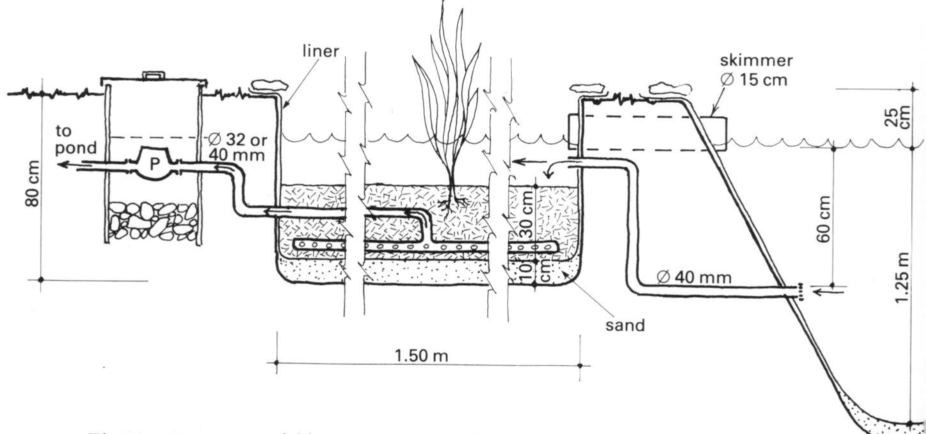

Fig 10 An external filter system can be built using the method described above. The filter can be constructed using concrete, fibreglass or a PVC or butyl-rubber liner. The water should be drawn off from the pond at about 60 cm (24 in) below the surface. This set-up could also incorporate a surface skimmer to remove floating debris, including leaves.

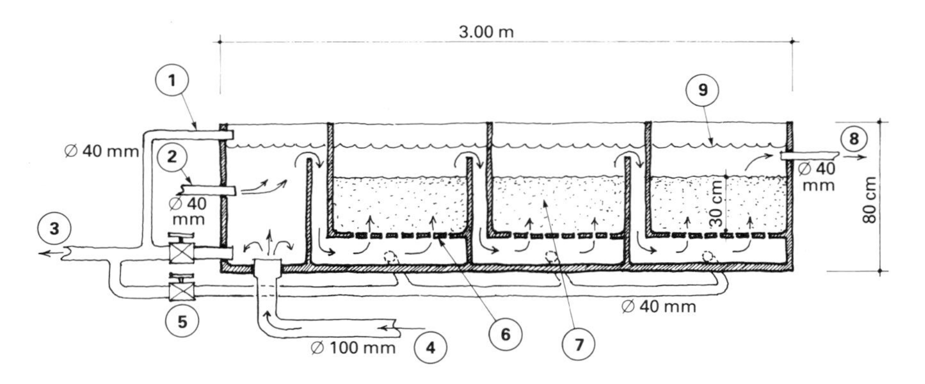

Fig 11 1 Overflow; 2 Middle feed; 3 To waste; 4 Feed from bottom drain; 5 To back-flush filter chambers; 6 Perforated plate; 7 Filter medium; 8 Return to pond via pump; 9 Pond water level.
Note: Before back-flushing, the bottom and middle feed should be closed. The bottom feed can be closed by means of a stand-pipe.

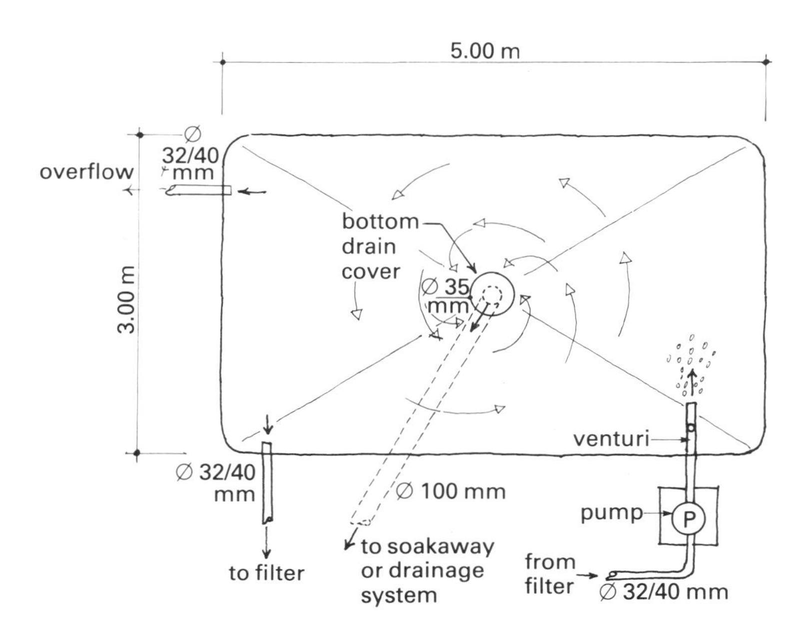

Fig 12 The water should circulate anti-clockwise.

cleaned regularly. The valve system incorporated when a sand filter is installed will enable you to clean out the filter by periodically back-flushing. Unfortunately, the binding agents used in some Koi pellet foods cause the sand to bind, impairing the efficiency of the filter. Therefore, if you install a sand filter, you should ensure that there is regular back-flushing and change the sand in the filter periodically to prevent malfunction.

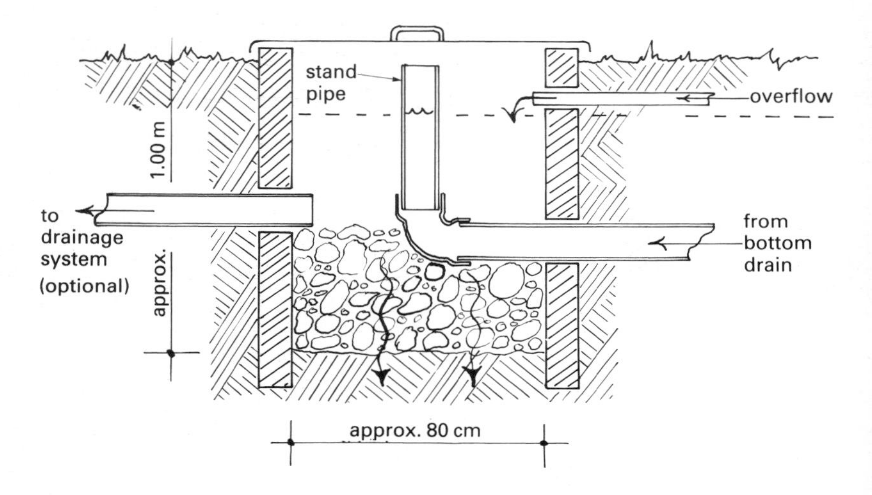

Fig 13 A discharge chamber incorporating a stand-pipe to operate a bottom drain. Quality PVC pipes and fittings with a reasonably tight fit should be used. A smear of grease or silicone paste around the bottom of the pipe will provide an excellent seal and allow you to remove the pipe periodically to operate the bottom drain.

Opposite: A waterfall is an imaginative creation requiring the delicate balancing of the water's force and counter force. It brings a pond to life and adds a soothing calmness to the surrounding atmosphere.

Plants for the Pond

Although it might seem obvious to add aquatic plants, this is not always the case for a pond containing Koi. Colour is provided by the fish themselves, and so you can dispense with plants altogether if you prefer.

Rockery plants and shrubs with less vibrant colours will provide complementary decoration for your Koi, and, together with miniature conifers, will create a definite oriental atmosphere. But if you prefer to add a variety of aquatic plants, without soil, so that you can clean the bottom of the pond regularly, they will need to be housed in perforated planting baskets.

A mixture of unfertilized potting soil and earth should be used in baskets with a minimum depth of 30cm (12in). To prevent the soil from escaping into the water, the baskets should be lined with hessian or a fine plastic mesh. There should be a liberal covering of washed gravel garnished with large round stones to prevent the Koi from disturbing the soil or uprooting the plants.

Marginal Plants

Most species of aquatic plant will need long periods of sunlight throughout the day. Baskets containing marginal plants can be placed on an appropriate shelf, or suspended over the sides of the pond at irregular intervals at the required depth. Alterna-

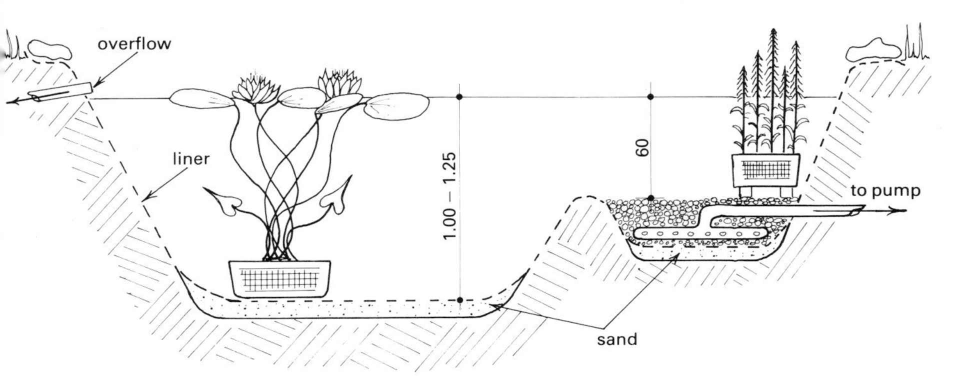

Fig 14 In a pond that has been made by using a liner, the basket should stand on a strong plastic sheet or an off-cut from the liner itself.

tively, to house marginal plants and for cleanliness, or for a pond that is not suitable for central plants, a bog garden containing the required depth of water can be made by digging a trough at the side of the pond. It will need to be 40–50 cm (16–20 in) deep and lined with a PVC or butyl-rubber liner. A thin layer of clean, soft sand can be spread on the bottom of the trough beforehand, to prevent sharp stones causing leaks. After laying the liner, fill the trough with a mixture of unfertilized potting soil and earth to a depth of approximately 25 cm (10 in). For a decorative finish, you could add a thin layer of washed gravel or stones.

When selecting plants, bear in mind their overall height, in order to keep your garden's landscape in perspective. Also, aim for beauty and continuity, choosing varieties of plants of various colours that will flower at different times of the year, as well as varieties that keep their flowers throughout the summer. To allow for growth, most marginal plants should be placed at irregular intervals of approximately 30–50 cm (12–20 in) and at a depth of 10–15 cm (4–6 in).

Water-lilies

The stronger, rustic varieties of water-lily are preferred. These will adapt to most climates. Their flowers can be red, white, yellow, pink or copper (orange), and each lily will need about 10

When we think of aquatic plants, most of us think almost immediately of the water-lily.

square metres (12 square yds) of water surface area. They will need to be planted at a depth of 60–125 cm (24–50 in) in perforated planting baskets with a depth of no less than 30 cm (12 in). After you have removed any dead leaves or roots, the rhizome, usually about 4–5 cm (1½–2 in) thick and from which the roots and leaves grow, should be planted at an angle between horizontal and 45°, just below the surface of the soil. A thin layer of clean gravel garnished with large stones will keep the soil and lily in place. The crown from which the leaves and flowers grow should remain exposed. A handle made from plastic-covered wire will help you to lower the basket into the pond and recover it for maintenance.

As with most other species of aquatic plant, success with water-lilies depends on early planting and on their constant exposure to the sun. Under these conditions, the flowers and

When you plant water-lilies, you can submerge any existing leaves below the surface of the water.

leaves will appear at the surface of the pond intermittently from late spring until mid-autumn, each flower lasting only a few days. It is not uncommon for there to be no flowers during the first year after planting.

Maintenance

There are very few diseases or predators of aquatic plants. Some of the more delicate submerged varieties may be nibbled by your fish, but this will not harm them. Koi of 70cm (28in) in length could uproot your water-lily, but if you have followed the planting instructions there is little risk of this.

Common pests such as greenfly might find their way from your roses to your water-lilies. Cut away and burn infested leaves and flowers and treat any nearby garden flowers similarly affected.

Note Do not use insecticides in or around the pond, since these are toxic to your Koi. Before treating the garden, check which way the wind is blowing.

To avoid pollution, the dead leaves and flowers should be removed periodically and completely cut back before the winter.

Rustic water-lilies and the majority of marginal plants will appear in the spring year after year and can be multiplied by simply dividing them at their roots. Lilies that have become particularly invasive will need to be reduced by dividing their rhizomes, leaving each section with its own crown and roots.

To avoid introducing parasites into your pond, wash the plants before planting in either a very strong salt solution or 20 mg of potassium permanganate per 5 litres of water. After washing, rinse the plants thoroughly in clean, fresh water.

Buying Koi

A common temptation for the new Koi enthusiast is to buy too many fish and so overstock the pond. However, the disastrous results soon teach the newcomer a valuable lesson.

One of the fascinating things about Koi are their vibrant colours and these fish are sometimes referred to in literature as 'living jewels'. For the beginner, it is difficult to come back from the pet shop, garden centre or dealer without a treasure. An experienced keeper will keep only a small number of good Koi and will know exactly what to look for when buying, bearing in mind the balance of colour in the pond. The Kohaku and Ogon are a very popular variety and easily found, so that the inexperienced tend to have too much red or gold in a Koi group.

Many patterns appear in the different varieties of Koi. It is a combination of the exactness of these with the intensity of colour, the body form, overall elegance, size and possibly an imposing appearance, that determines the value of a Koi.

When choosing baby Koi, you should be aware that their patterns may not possess the same growth rate as their bodies. A small Koi with a perfect pattern might not be so impressive as it gets older, while a baby Koi with a large, monotonous pattern could grow into a beauty. Black and red patterns on certain varieties will at first appear under the surface of the skin, but will become prominent as the Koi grows up. This tendency is often seen when they have recently been taken from their natural environment.

Regrettably, there are some stockists who keep Koi under unfavourable conditions – in the worst cases, in aquarium tanks outside the shop. The subsequent rapid temperature changes undergone day and night by the Koi result in stress, lowering their resistance to disease. Eventually, they either die or are sold in poor condition, possibly leaving behind for the new arrivals a disease that will take over as their natural resistance in turn is weakened.

A good dealer will do his best to stock Koi under optimum conditions in order to sell healthy fish. Such a dealer will also get to know you and try to fulfil your requirements, giving you help and advice along the way.

There can be no guarantee that a Koi will not become sick. Once you have bought the fish, it is entirely your responsibility to provide ideal living conditions to prevent health problems. Healthy Koi are bright-eyed, active and hungry, with a normal breathing rate. The dorsal fin will be erect and the tail and pectoral fins will be free from erosion. Split fins will repair themselves. There should be no reddened areas on the body or at the tail or fin joints. The body should not appear thin, but must have a smooth, solid outline. Koi destined to grow large will have a round body that is thick at the tail joint.

Opposite: Trying to predict how a young Koi's pattern will turn out is an exciting and intriguing part of choosing fish.

Transporting Koi

Travelling and a change of environment can prove extremely stressful to Koi, and so certain precautions should be taken. They should travel in a plastic bag containing just enough water to allow them to move around, which should be placed inside a second bag to avoid mishaps. The dissolved oxygen used will be replaced by the air in the bag. For a journey of several hours, use the same method, but fill the inner bag with pure oxygen before sealing it. To keep the fish calm and the water temperature stable, place the bag inside a cardboard box.

Before introducing Koi into your pond, you must float the bag they have travelled in on the surface, avoiding direct sunlight, to allow the temperature of the water in the bag to adjust itself slowly to that of the pond. A difference of more than 5°C (9°F) can prove harmful, so, for safety, allow a minimum of 20 minutes before introducing them into the pond. Your Koi group can be fed after about an hour, which could help to establish the timid newcomer who might decide to hide for a few days.

Opposite: Koi should always be transported with great care.

Varieties of Koi

All varieties of Nishikigoi or Koi, have been produced to be viewed from above and are all from the same species of carp, with the exception of the Doitsu (scaleless type), which are the results of interbreeding leather carp and mirror carp.

Each Koi's colours and patterns are different, just as the tastes of each keeper are different. The advice below will help you to choose the Koi you prefer. The permutation of pattern and colour is almost limitless, so remember, if you see a good, healthy Koi that appeals to you, it does not have to look exactly like the photographs in this book. If you look hard enough, you will always find what you are looking for.

Kohaku

The white texture of the Kohaku's body, tail and fins should be as pure as possible. The red markings, known as *hi* (*hee*), must be intense and distributed evenly along the body and above the lateral line, creating a well-balanced design. Large *hi* patterns are more impressive and should start on the head and be distributed evenly to the tail, leaving the end of the tail joint white.

The head pattern must reach almost to the mouth, but must not cover completely the eyes or gill covers. Although sometimes, as with other varieties, an unusual marking can impart great charm or beauty, a pattern that does not descend below

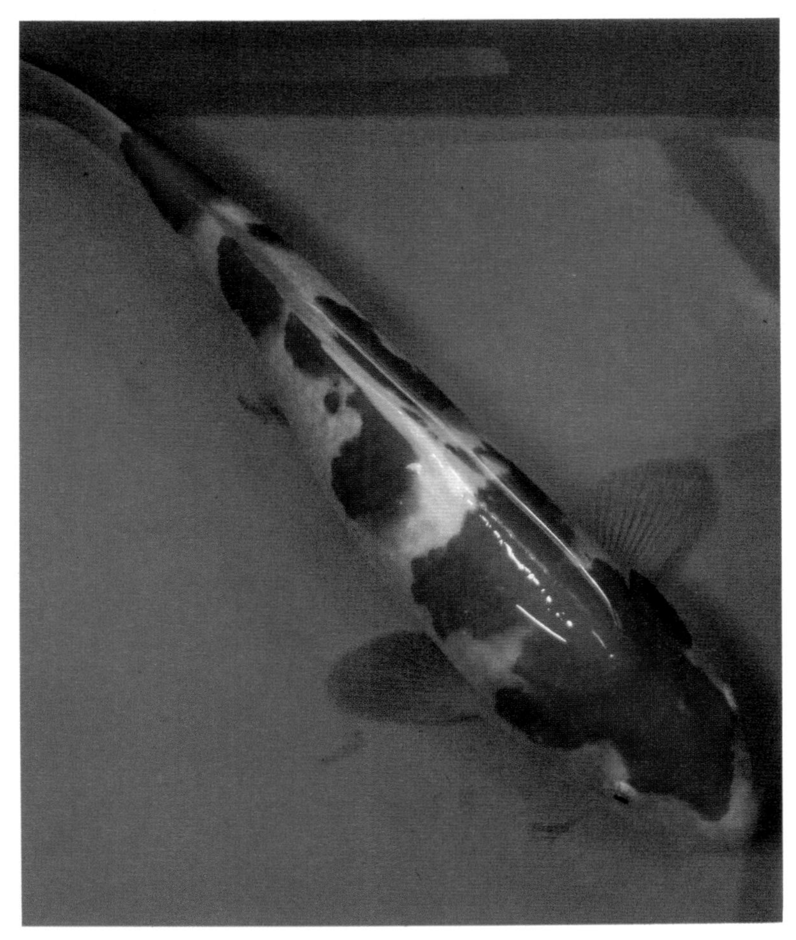

Displaying the national colours of Japan, the Kohaku (*ko-ha-ku*), with its red and white markings, is the most popular of all the varieties of Koi. The Tancho Kohaku is a white Koi with a red crown on its head. The name derives from the Tancho crane, which is known in Japan as the 'bird of happiness' and is regarded as a symbol of love. It, too, has a red crown on its head (see *Fig* 15, p. 62).

the lateral line or cover the mouth, eyes or tail joint, will provide a balance and allow the body fins and tail to set off the pattern to its maximum advantage.

Fig 15 The Tancho crane.

Taisho Sanke

The Sanke (*tay-sho-san-kee*) is a white Koi with red and black patterns on the body. Elegance can be found through a large *hi* pattern, similar to that of the Kohaku, with a smaller black pattern, known as *sumi* (*sue-mee*), in between. A few *sumi* stripes on the tail or fins can enhance the overall beauty.

The *sumi* pattern can also be on, or mixed with, that of the *hi*, but this can give the overall design a heavy appearance, especially if it appears on the head. Each pattern should be distributed as evenly as possible above the lateral line and that of the *sumi* should reach only as far as the shoulders.

Bear in mind, when buying a baby Sanke, that one with very few *sumi* markings but a large *hi* pattern will possibly become more elegant as it grows.

Large *sumi* patterns on a baby Koi will possibly break up with age, whereas *hi* patterns usually remain stable. Sometimes, however, a small *hi* pattern may appear to shrink in proportion with the body of the Koi as it grows.

A white Koi with a red crown on its head and a *sumi* pattern on the body is called a Tancho Sanke.

Showa Sanshoku

The Showa Sanshoku (*showa-san-show-coo*) is a black Koi with red and white patterns and was bred a few years after the Sanke. The Showa is very similar, being red, white and black, but is distinguished by having *sumi* and *hi* as the dominant colour. Again for elegance, as with the Kohaku, a large *hi* pattern is preferred, but this time with a large *sumi* pattern that may descend below the lateral line and also appear on the head. The fins and tail should be white, but the pectoral fins must have *sumi* at the joints. There is sometimes a noticeable difference in the texture of the *sumi*, and on some Showas it will assume a purple hue.

A Showa with more white, or an equal amount of white, is called a Kindai Showa, and where *hi* is the dominant feature from the head to the tail, it is known as a Hi Showa. A Showa lacking a *hi* pattern, but with a red crown on its head, is known as a Tancho Showa.

The Taisho Sanke was created at the turn of the century and was the first tricoloured Koi to become established.

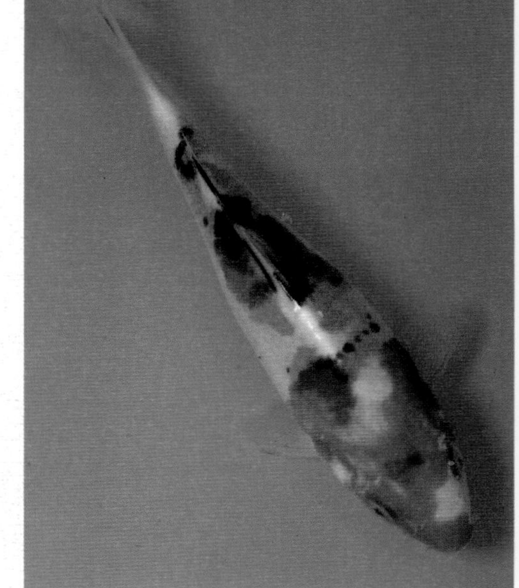

Sumi patterns can sometimes be seen under the Koi's skin and may appear on its surface at a later stage, depending on several factors, including diet and water conditions.

A Sanke with a metallic hue is known as a Yamatonishiki, while one with an unbroken red pattern from head to tail is called an Aka Sanke.

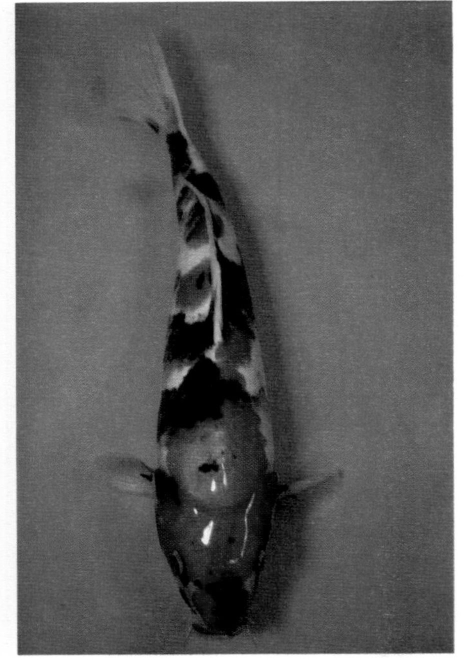

In proportion to the white of the fins and body, the *hi* and *sumi* patterns will be the dominant feature of a Showa.

Ki Utsuri

The Ki Utsuri (*kee-oot-sue-ree*) have black as their predominant skin colour and are said to be partly responsible for the creation of the Showa Sanshoku. Utsuri Koi can have red *hi,* white *shiro* (*shearo*), or yellow *ki* (*kee*) patterns. The texture of these should be as pure as possible and distributed evenly from the head to the tail and should flow over the body and below the lateral line. The tips of the fins or tail can be white, revealing the similarity with the Showa.

Doitsu

Doitsu (*doyt-sue*) Koi are the results of interbreeding the German leather carp and the mirror carp. (Doitsu is Japanese for 'German'.) Apart from a single row of scales each side of the dorsal fin, the leather carp is scaleless. The mirror carp has a single line of larger scales that run from the shoulders to the tail, following the lateral line and across the length of the back, dividing into two rows each side of the dorsal fin.

Elegance comes not only from the colours and patterns, but also from the size and regularity of the lines of scales themselves.

Shusui

Shusui (*shoe-suey*) Koi have a clear blue head and a light-blue back with a perfectly arranged line of dark-blue scales that run along the lateral line, over the shoulders and along the back to the tail. The abdomen, tip of the nose, fin joints and gill covers should be dark orange. When the orange appears above the lateral line, the fish is called a Hana Shusui, and if the back is completely covered, it is known as a Hi Shusui.

The Utsuri Koi that have an orange or red pattern are defined as Hi Utsuri.

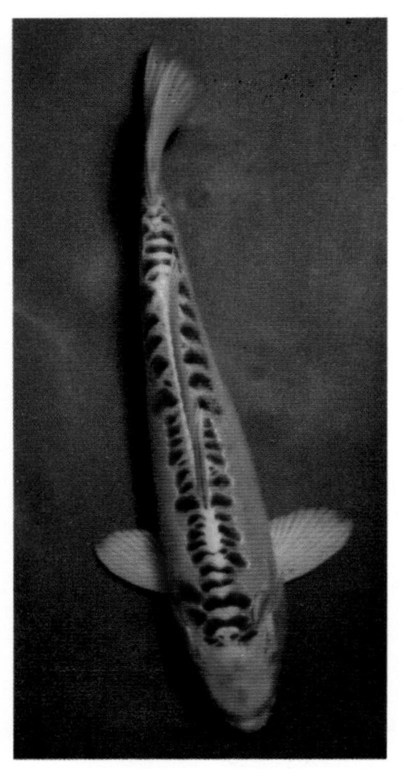

Matsuba Ogon Doitsu.

Older fish of the Doitsu class will sometimes have scales missing, particularly along the lateral line. This loss is probably caused by the fish rubbing itself, due to irritation, on the bottom or sides of the pond, or by handling. An example of this can be seen with the German mirror carp (Doitsugoi) illustrated.

Hajiro Doitsu. Any Koi with this type of scale distribution may be classed as a Doitsu. The term is sometimes used before or after the name of the variety: Doitsu Kohaku or Ki Utsuri Doitsu, for example.

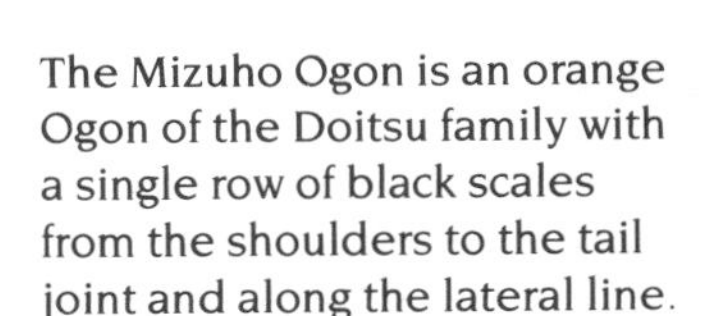

The Mizuho Ogon is an orange Ogon of the Doitsu family with a single row of black scales from the shoulders to the tail joint and along the lateral line.

Although one of the first representatives of the Doitsu class, the Shusui is also an exception in not taking the additional name Doitsu. The name Shusui both pays homage to its creator and suggests autumn.

Kin Matsuba (*kin-mat-sue-ba*) or Matsuba Ogon is the name given to a golden Koi which has highly embossed scales.

The golden Koi, known as an Ogon, and the silver Koi, known as the Platinum Ogon, took many Koi generations to create, becoming established just after World War 2. This introduction made a great change to the already established varieties by adding a gold (*kin*) or silver (*gin*) (hard 'g') metallic finish. For example, a Matsuba Koi with a platinum texture is called a Gin Matsuba. The Ogon is partly responsible for attracting many newcomers to the 'Art' of Koi keeping.

Hariwake

Hariwake is the name given to a Koi with a gold and silver pattern.

Kin, Gin Rin

All varieties of Koi can exhibit the highly metallic scales known as Gin Rin. Some of these scales look as though they have been brushed and are known as Hiroshimanishiki. Scales that assume a golden hue are referred to as Kin Rin. The names Kin, Gin and Rin can be placed before or after the name of the particular variety, or even divided. for example: Gin Rin Kohaku, Shiro Bekko Gin Rin, Kin Ki Utsuri, etc.

Highly metallic scales that appear on a red texture are called Kin Rin.

Shiro Bekko Gin Rin

Bekko Koi have an overall skin colour of red (*aka*), white (*shiro*), or yellow (*ki*) which again explains that when the red pattern is continuous from the head to the tail and unbroken by any other colour it can be called *aka*. Although similar, the noticeable difference between the Utsuri Koi and the Bekko is that the *sumi* is confined to the back and does not descend to the abdomen below the lateral line. A good specimen must have little or no *sumi* markings on the head, but they must, at the least, reach to the shoulders. These similarities reveal that Bekko Koi are related to the Sanke family.

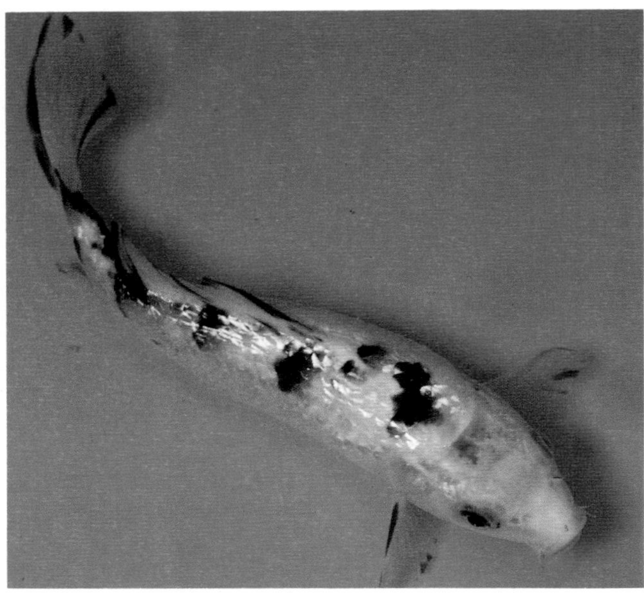

This Tancho Kohaku has a *sumi* pattern that can be seen through the skin. Under the right water conditions and with correct feeding this Koi will perhaps become an excellent Tancho Sanke.

The intensity of colour of your Koi will vary with changes in the light throughout the day. More noticeably, their colours will be at their best during the autumn. A varied diet throughout the year of pellet foods containing small amounts of carotene and spirulina, combined with natural foods, will help to maintain the colours of your Koi to their best advantage.

Exhibition-class Koi can be ordered from your local dealer, who will obtain photographs from which you can make your choice. The features to look for in fish to be exhibited are as described. For more details, visit a local Koi show and join one of the associations or clubs.

The standards and rules for judging are very similar throughout the various Koi clubs and associations. In general, although the patterns on some Koi can be the main attraction or even make a fish a show stopper at first glance, the judges will also consider the Koi's figure, colour, elegance, quality or imposing appearance.

Breeders in Japan work hard on improving the different colours and varieties of Nishikigoi. This progress, together with new colours and patterns that arrive as the parentage of Koi changes over the years, is one of the principal factors that make Koi an interesting and rewarding hobby.

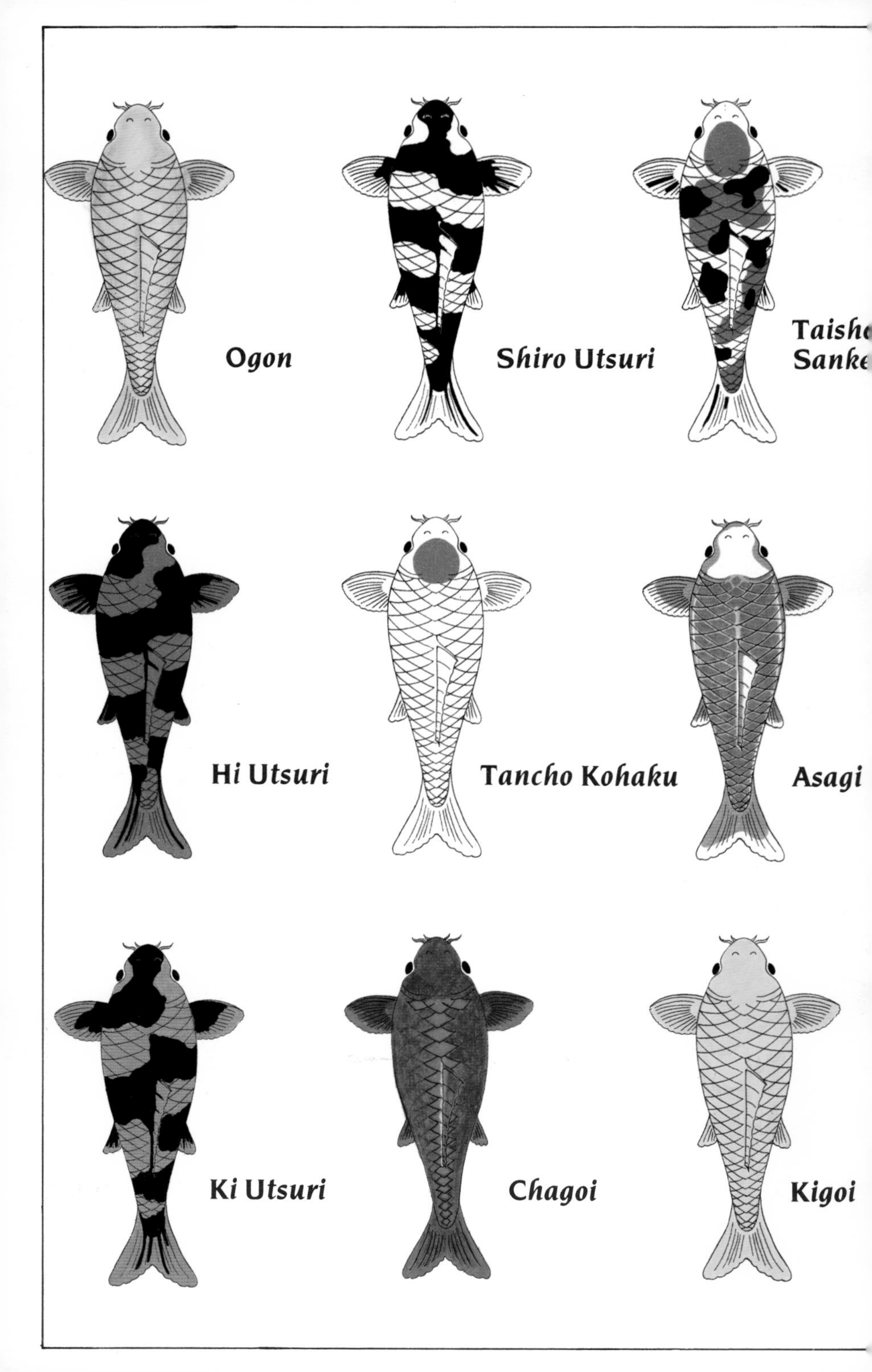

Ogon
Shiro Utsuri
Taisho
Sanke
Hi Utsuri
Tancho Kohaku
Asagi
Ki Utsuri
Chagoi
Kigoi

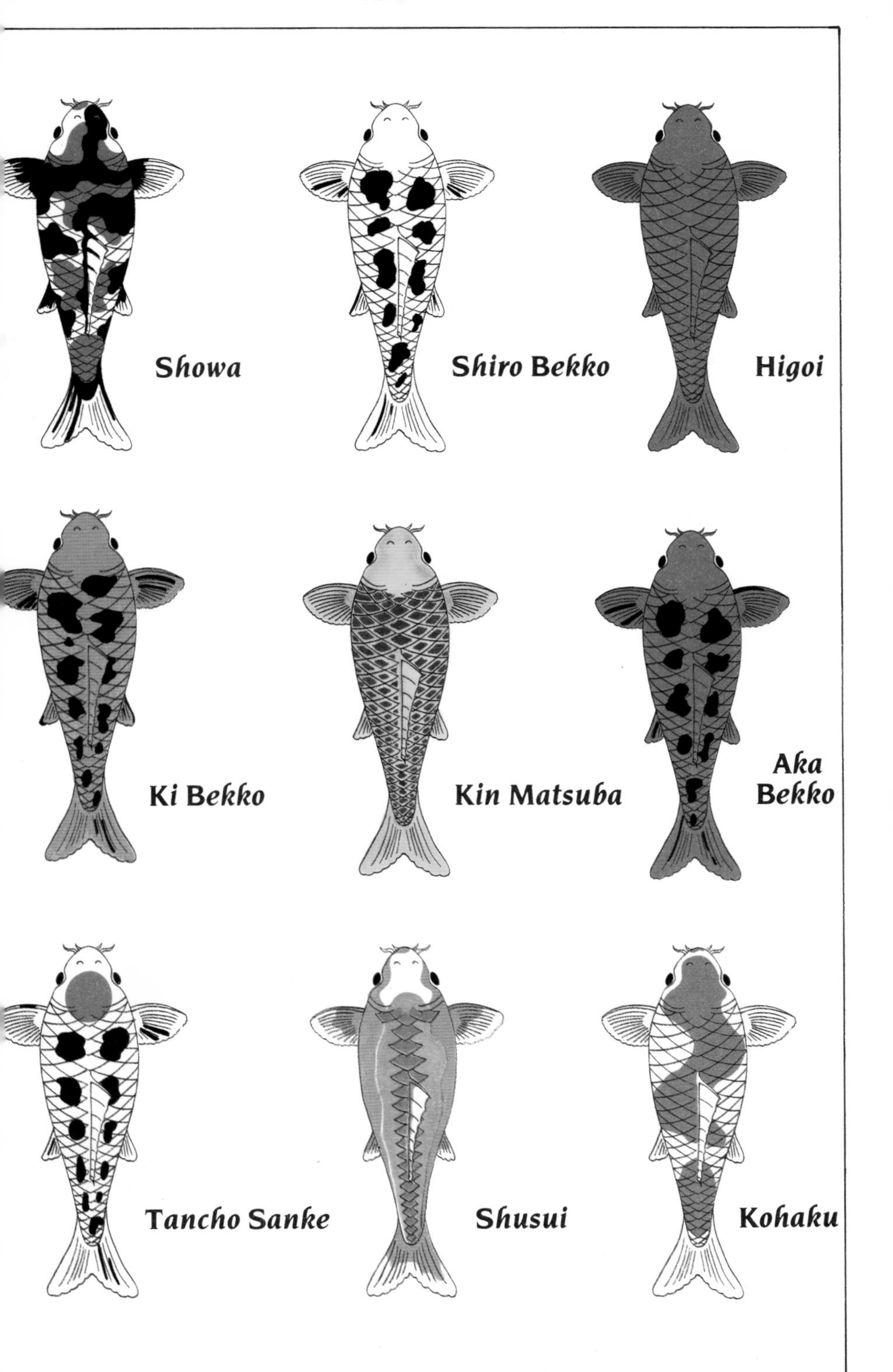
Showa
Shiro Bekko
Higoi
Ki Bekko
Kin Matsuba
Aka
Bekko
Tancho Sanke
Shusui
Kohaku

Morphology

Koi are descendants of the common carp (*Cyprinus carpio*).

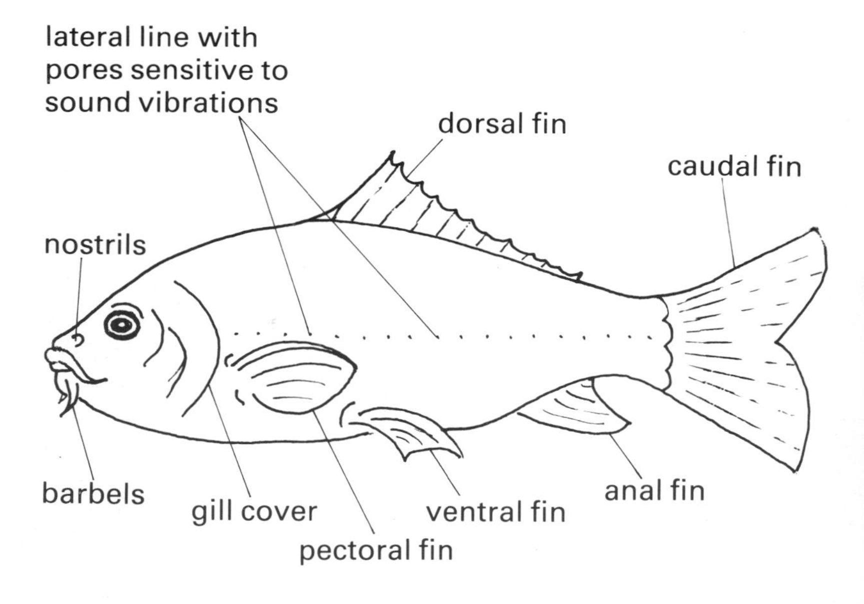

Fig 16 Some physical characteristics of Koi.

Gill covers

Behind the gill covers are the red gill filaments that transfer oxygen to the red blood cells. In turn, the red blood cells are transferred by the heart to the rest of the body. In order for the fish to obtain enough dissolved oxygen, the gill covers have to move large quantities of water over the gill filaments.

Nostrils

The nostrils contain sensors that react in a similar way to the taste buds on the tongue of a mammal. Enabling the Koi to perceive odour, the sensors are connected to the brain via the nervous system.

Barbels

There are four barbels, two each side of the Koi's mouth. These are used to detect food below the surface while the fish is feeding on the bottom and are sometimes referred to as the Koi's moustache.

The dorsal and anal fins are used to stabilize movement, while the body and caudal fin propel the Koi. When watching Koi swimming, you can clearly see the whole body and caudal fin used as a rudder and the ventral and pectoral fins acting as ailerons to change depth and to brake.

Growth Rate

The growth rate of your Koi will depend entirely on their environment, which is influenced by pond depth, stocking rates, filtration, water temperature and feeding.

Koi grow at different rates and it can be difficult, in some cases, to know their exact age. As with humans, their growth rate slows down as they reach adulthood. For an approximate guide, a Koi can grow to 10–20 cm (4–8 in) in length during the first year and 20–30 cm (8–12 in) in the second. Between the ages of two and three years, they will become adult and possibly grow a further 10 cm (4 in), but it could take another seven years to reach the average length of 70 cm (28 in).

Koi have a long life span compared with other ornamental fish and, depending on their environment, will live to 15–20 years or even more and reach a weight of 5–10 kg (11–22 lb). The above figures are only an approximate guide to age. The scales of a Koi will reveal its true age since they are marked by growth rings that can be seen if a scale is examined under a microscope.

The greatest recorded age attained by a Koi is more than 200 years.

It is often difficult to determine the exact age of Koi.

Environment

Koi prefer freshwater with temperatures in the range of 8°–28°C (46°–82°F). Below these temperatures they will remain at the bottom of the pond. In winter, when the temperature outside the pond can fall below freezing, Koi will prefer to remain sheltered in the deepest part of the pond, where the water temperature will remain around 3°–4°C (37°–39°F). If it gets colder, Koi will die.

Out of their natural environment, Koi need clean, filtered water with plenty of oxygen and a pH level of around 7.5.

pH is a scientific but generally useful term used to express the acidity or alkalinity of any substance. The pH scale goes from 0 (acid) to 14 (alkaline), 7 being the neutral position. Kits for testing pH are simple to use and can be bought at a pet store, garden centre or from a Koi dealer. To make a test and ensure that your filter is working efficiently, the water sample should be taken from the average depth of the pond, which will give you the most accurate reading. Koi will suffer if kept for long periods in water lacking the recommended pH level.

Feeding Koi

Feeding should be considered as essential a part of Koi keeping as controlling the quality of the water. The most common mistake is to give too much food at a time. The uneaten food and fish waste, if left to accumulate, will soon upset the balance and quality of the pond water, which will in turn have adverse effects on the health of the fish.

Generally, Koi derive benefit from only a small proportion of the food they are given. The rest will pass through their digestive tube and end up on the bottom of the pond as waste.

The beginner is advised to feed his Koi twice a day (morning and early afternoon), giving them only as much food as they can eat in a few minutes. Any food left uneaten should be netted until you have been able to calculate exactly the amount of food required for your particular Koi group.

They will benefit more from their food as the water temperature increases. The amount distributed should be gradually increased from spring until the height of summer, and then reduced in autumn as the water temperature declines.

Keep a thermometer beside your pond to monitor the changes in water temperature. These will vary slightly from region to region and according to the depth of your pond, but the water will be at approximately 8°–10°C (46°–50°F) in spring. Feeding should be kept to a minimum until the water temperature starts to rise, and then you should gradually increase the amount of food accordingly.

Koi will eat as much food as you give them, but will benefit from only a small proportion of it.

At high-summer temperatures, you should feed Koi generously, increasing feeding from two to three times a day: morning, early afternoon and evening. This is the time when they will grow and build up the surplus energy that will help them through the winter. Where possible, the filtration and oxygen supply should be increased and the bottom of the pond cleaned more frequently.

If you take a summer holiday, you can leave your Koi without food for a few weeks without them suffering any ill effects. In fact, the break will be beneficial to them, too.

If you plan to stay away longer, ask a trustworthy friend or neighbour to feed them, giving strict instructions and providing ready-prepared packets of food or, more simply, a measure, to ensure that the right amount of food is given. In certain cases, depending on your garden's situation, it may be advisable to cover part of the pond to provide shelter for your Koi and protect them from any fish-eating predators, who may take advantage of the fact that you are not there.

If you have a particularly large collection of Koi, it is also advisable to instruct your friend or neighbour on how to clean the bottom of the pond. This should be done at least twice a week during your absence.

Autumn

As the water temperature starts to drop, during early autumn, reduce feeding times to twice a day, dispensing with the early-evening feed. Monitor the water temperature and decrease the amount of food given accordingly. At this time you should gradually change your Koi's diet from a food rich in animal protein to a vegetable-protein type. The vegetable-protein-rich food is easily digested at low water temperatures. Again, as the water temperature approaches 10°C (50°F), feeding should be reduced to a minimum, and at below 8°C (46°F) it is advisable to cease feeding altogether. Any food given at such low temperatures would be poorly digested, which could cause problems. At this time your Koi will be surviving on energy stored during the summer.

Types of Food

Many prepared foods are available nowadays. The beginner should feed their Koi only on the floating pellet types specially prepared for pond fish, to ensure that they receive a well-balanced diet. There are two basic Koi pellet foods: the animal-protein type (staple) and the vegetable-protein type (wheatgerm). Easily digested, the wheatgerm type is specially prepared for lower water temperatures, while the staple type ensures energy and growth during the summer. Pellets come in various sizes; baby Koi should be fed on the smaller size.

Floating foods enable you to admire your Koi while they are feeding and, if necessary, the uneaten pellets can easily be netted. Flake foods, by contrast, are less entertaining and tend to foul the water.

Live foods can be fed to your Koi, but if they are obtained from an unreliable source they can introduce parasites and disease into your pond. If you are a beginner, feed only the floating pellet types of food until you find a dependable source of cultured live food that will enable you to vary your Koi's diet. Generally, aquatic insects will start to multiply in your pond and these will also be enjoyed by your fish.

Training Koi

Koi keeping is a pleasurable pastime, bringing tranquillity to any garden. A further pleasure is to be able to feed Koi from your hand. Careful feeding and regular feeding times are very important while training your fish, and a set pattern with slow movements will avoid startling them as you approach the pond.

A basic method is to stop feeding altogether and, for a few days, appear slowly each morning at the ponds' side, avoiding sudden movements. On the third morning, reach down slowly and gently touch the water with your hand. This should be repeated each morning for a further two or three days and finally, on the sixth or seventh morning, you should leave a little food. Now hungry, the Koi will associate your approach with feeding time and will subsequently greet you each time you arrive.

When there are just a few Koi, training becomes a little more difficult. However, the method can be repeated if you are not successful at the first attempt. Always feed your Koi in the same part of the pond each time.

When there are many Koi in a pond, they are usually trained more easily.

Without any doubt, one of the most enjoyable moments during the day is watching your Koi feed.

Winter until Spring

Winter and spring are very important times of the year for Koi keepers. An extra effort should be made to ensure that your Koi enter their hibernation period in peak condition, retaining their resistance to low winter temperatures and to the disease and parasites that will abound the following spring.

Sometime in late autumn, dead leaves and waste should be removed from the bottom, and plants, if any, should be cut back in and around the pond. A treatment against parasites will help reduce the numbers of those that will multiply when spring arrives.

Filtration should be maintained throughout the winter, but since the water at the bottom of the pond stays warmer than that at the surface, any current produced by the filter return or the venturi should be reduced, to avoid cooling the bottom water or disturbing the Koi.

Generally, a pond cover will help to stop ice forming, but this depends on the surrounding edge of the pond being able to accommodate a cover and, of course, on your location. A pond cover will trap a layer of air over the surface of the pond that will help to prevent ice forming and also stop any leaves from falling into the water. A cover can be made from fine-mesh greenhouse shading available at a garden centre. This has an advantage over a plastic sheet in that heavy quantities of rainwater cannot collect on its surface. It can be stretched over a wooden frame for easy handling and, if the pond is large, several frames can be placed side by side.

A gentle current will help keep part of the surface free from ice and so allow an adequate supply of oxygen to reach the water.

There are many methods of keeping the pond's surface free from ice, but a gentle current is probably the best. However, if ice should cover the whole pond, a recommended remedy is to cut, drill or melt a hole approximately 5 cm (2 in) in diameter and syphon off about 15 cm (6 in) of surface water. This should prevent further freezing by creating a pocket of air between the ice and the surface of the pond. Several such holes at different places around the pond will allow an adequate oxygen supply to the water. To prevent cold air entering, the holes should be covered with hessian cloth.

Holes in the ice should be made only by cutting, drilling or melting with hot water. Never use a hammer or a pick, since their vibrations, transmitted through the ice and water, could have adverse effects on the Koi or possibly on the structure of the pond itself.

Spring

As the rising spring temperatures warm your pond, the Koi will be seen looking for food. At first their activity will be slow, and so they should be fed sparingly. If the pond has been covered, the cover should be removed in stages over a period of a week, to avoid a sudden, stressful change in the Koi's environment.

A treatment against parasites that are already established and multiplying in your pond will help reduce their numbers, giving any Koi that have been weakened by the stress of winter a chance to gain their strength. The fish should be observed closely for any signs of disease or parasite attacks. Skilful observation and quick action (see p. 95) may well save any distressed Koi.

If your pond receives adequate sunlight, the water may turn green with the growth of algae until your filter builds up enough helpful bacteria to deplete their numbers. If necessary, partial water changes, over, say, two weeks, will help to clear the water. Refill your pond with clean tap water directed onto the surface through a spray. This will help to vaporize any chlorine that might be present as a purifying agent in your water supply. Never use chemicals to clear your pond water of algae, for although they might not be harmful for your Koi, they will possibly destroy the useful bacteria established in your filter, rendering it inefficient and thus allowing the accumulation of toxic levels of ammonia and harmful nitrites.

Those wise enough to have installed a variable-speed central-heating circulator will now be able to increase its flow rate, and so accelerate filtration. The extra water current will increase aeration and also help your Koi to develop healthy figures and appetites.

Summer

When your pond water reaches approximately 15°C (59°F), you will be able to change your Koi's diet gradually from a vegetable-protein to an animal-protein type. As the feeding rates and your Koi's activities increase, there is an even greater need for regular cleaning, good filtration and aeration of the water.

Heavy feeding and warm weather will rapidly deplete the oxygen dissolved in water, and at times during the summer the weather may become heavy and overcast, particularly a few days before a storm. Such weather has a remarkable effect on the dissolved oxygen level, reducing it to a minimum within hours. At such times a waterfall or fountain should be left flowing throughout the night.

In summer aquatic life is at the height of its activities. This includes not only your Koi, but also the various organisms that are present in every pond. The breeding cycle of these varies from type to type, and the particular organisms that can harm your Koi can be eliminated by regular treatment of your pond. This remedy is explained in the following chapter.

Opposite: In summer Koi will grow and store the surplus energy that will help them survive the winter.

Diseases and Treatments

It is difficult in some cases to cure sick fish, so remember that prevention is better than cure. Stress, pond depth, insufficient filtration, overcrowding, lack of oxygen, excessive feeding and the accumulation of toxic waste, are the main problem areas.

If, despite all your efforts, you do have a problem with disease, this chapter will enable you to recognize and understand any adverse symptoms, as well as help you to provide a cure. It is important to understand the origin and cause of any illness and to identify and treat it at an early stage. Familiarity with your Koi and their behaviour patterns will help you in this respect.

In their natural environment, fish are able to swim about freely in large volumes of water, selecting their own diet. The accumulation of their metabolic waste cannot reach a level high enough to affect their health. Harmful organisms will feed on wastes and fish, but are unable to reach sufficient numbers to affect the health of the fish.

In an artificial environment, however, if conditions become unsatisfactory through overcrowding, overfeeding or the accumulation of solid and dissolved wastes, these harmful organisms can reproduce easily and reach such high levels that they overcome the natural resistance of your fish.

Koi's main means of resistance to organisms such as parasites, bacteria and fungi is a protective mucus coating over their skin and scales. This will ward off these organisms if their levels are not too high and as long as the coating remains in good condition and undamaged. Because of the function of this mucus coating and its vulnerability, it is very important to maintain optimum conditions and exercise great care should it be absolutely necessary to remove a Koi for treatment.

Diagnosis

Generally, a sick Koi will leave its group, which in itself may well be the first sign of illness. Other signs include loss of appetite, swimming in a listless manner with all the fins closed, or gasping for oxygen at the surface. Do not jump to conclusions if one of your Koi does not rise to eat or decides to go off hunting alone at the other end of the pond. Patient and careful observation will help you to recognize any real problems.

Many types of disease can affect Koi. To simplify the diagnosis and cause of those most likely to be encountered, we can divide them into three main groups: parasite infestation, bacterial infection and fungal growth.

Parasites

Parasites have many ways of finding their way into your pond, through passing birds, in live foods or possibly via a new member of your Koi group. Most of the parasites will be ecto-parasites, those that live on the outside of the fish. It is usually the weakest member of the group or one with a damaged mucus coating that will become the host. Some parasites can be seen with the naked eye, while others can be seen only under a microscope. Nearly all need a host fish on which to survive and reproduce. Usually, eggs are released into the pond, hatching after a variable period, the length of which depends on the water temperature.

In winter, the eggs and parasites will lie dormant, like the fish themselves. As water temperatures rise in the spring, the eggs will hatch and the parasites will start to look for a host. Generally, the fish that have suffered the most from the stress of the winter temperatures will be chosen. Treating the pond for parasites (see p. 110) at the beginning of spring will help to reduce their numbers and enable the fish to regain their strength

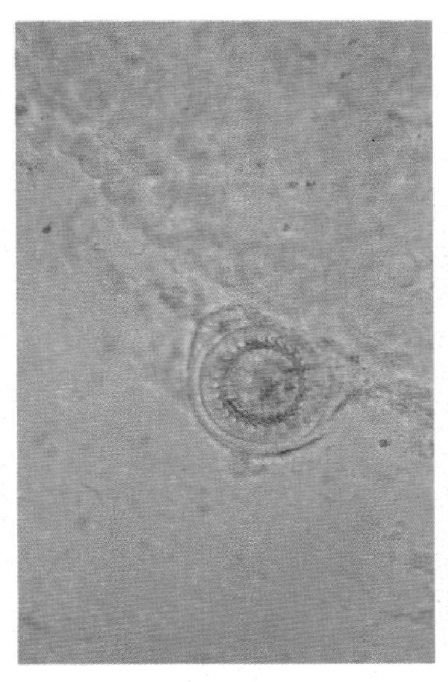

Trichodina on skin. Diameter approximately 50 μm (50 microns).

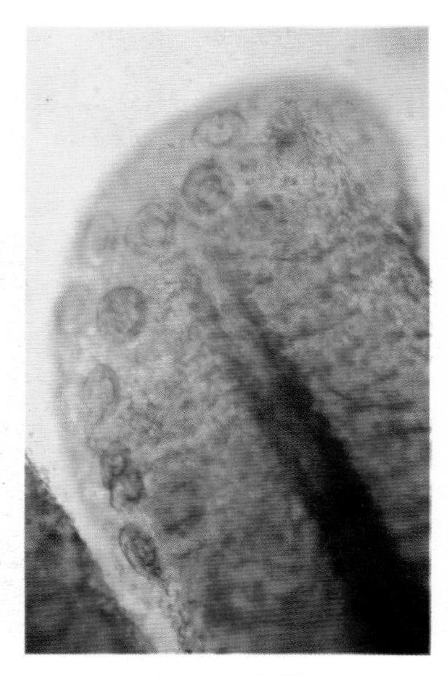

Trichodina on gill filament.

Chilodonella on skin. Length approximately 500 μm.

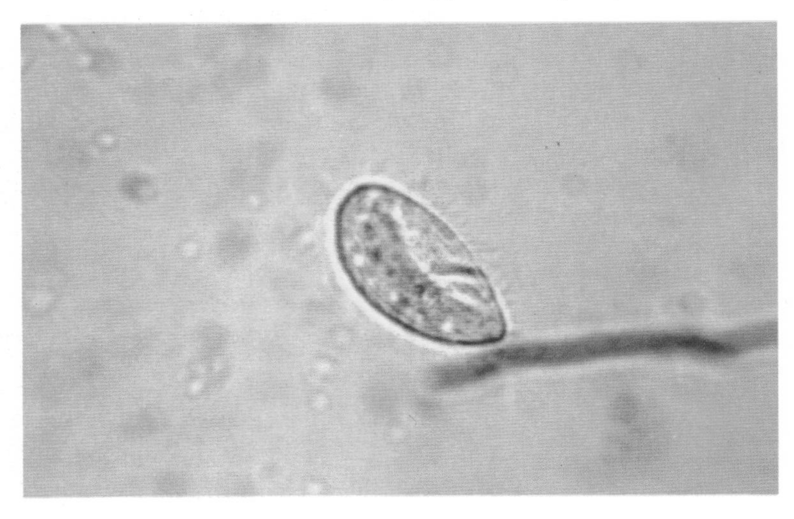

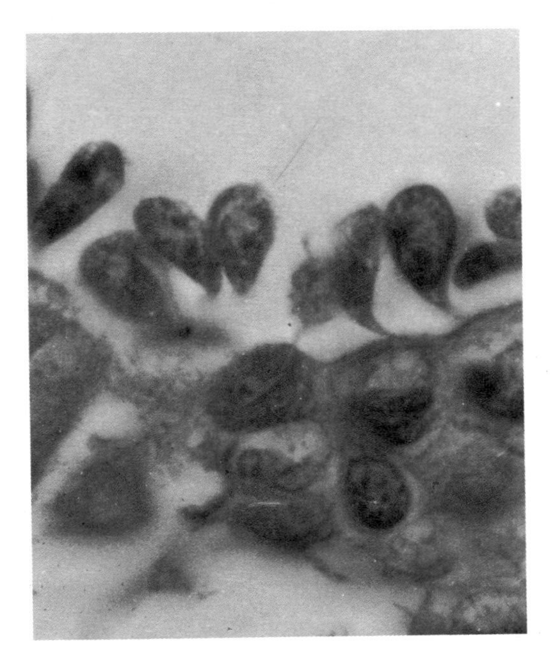

Costia on skin 10 μm (x 1100).

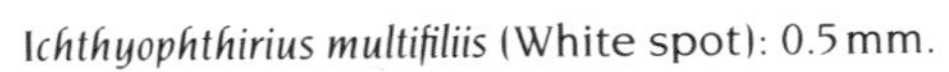

Ichthyophthirius multifiliis (White spot): 0.5 mm.

and resistance, before the parasites can cause any damage.

Where there are fish, there will always be fish parasites. Some feed on organic debris in the pond and will use the fish as their transport. Cleanliness will help to control these, but other types attach themselves permanently to the fish, feeding on the body tissues and fluids.

As in their natural environment, healthy fish are able to support a small number of parasites without suffering any ill effects. In any case the parasites will be limited in number, since not all of them will find a host on which to survive. In an artificial environment, however, where the fish live in a restricted area, parasites have more chance of finding a host and can easily transfer from fish to fish. Under these conditions, they are also able to multiply quickly.

The danger of parasites lies in the damage they can cause to the skin and fins or gills of the fish, which is unable to heal quickly enough before harmful bacteria or fungus infects the site of attack.

Protozoan parasites

The most common of all parasites that can affect Koi are of the single-celled external protozoan type. These live on the skin or gills and can be seen only under a microscope. Unusual behaviour by your Koi may reveal the presence of one or more types of such parasites. An infested Koi will have its fins, specially the dorsal fin, closed and will spend much of its time alone appearing listless and possibly lacking appetite. As a result of the irritating effects of these parasites, Koi will also be seen rubbing themselves on the bottom or sides of the pond. While doing so, they will turn on their sides, exposing their lighter abdomens to the light, in what is commonly known as 'flashing'. In some cases, the irritation caused by certain parasites will make the Koi produce extra mucus, which can appear as a blue-grey film on the skin. If the gill filaments become infested, the Koi can be seen 'flashing' to free itself of the parasites, breathing rapidly or gasping at the surface of the pond in order to secure more oxygen.

The photographs (see pp. 96, 97) show some of the protozoan parasites that can infest Koi. Fortunately, they can all be controlled by the same type of treatment.

Gyrodactylus

Two other parasites that can affect Koi come from the monogenetic trematode family. The first is *Gyrodactylus*, commonly referred to as 'worms' or 'flukes'. These attach themselves to the Koi by means of hooks in order to feed, and although rather small, can just be seen with the naked eye. However, to confirm the diagnosis, they should be identified with the aid of a magnifying glass or a low-power microscope.

During its infestation of Koi, *Gyrodactylus* will eat away the inner rays or all of the tail or fins, and continue on reaching the skin. At this stage the fish may well die through stress or a bacterial infection at the sites of attack. Once the parasite has taken hold, the course of destruction is quite rapid, and so tail and fin inspection, as with various other diseases, should become a habit while you admire your collection. *Gyrodactylus*, being viviparous, produce their young fully developed one at a time.

Gyrodactylus at work. Length approximately 700 μm.

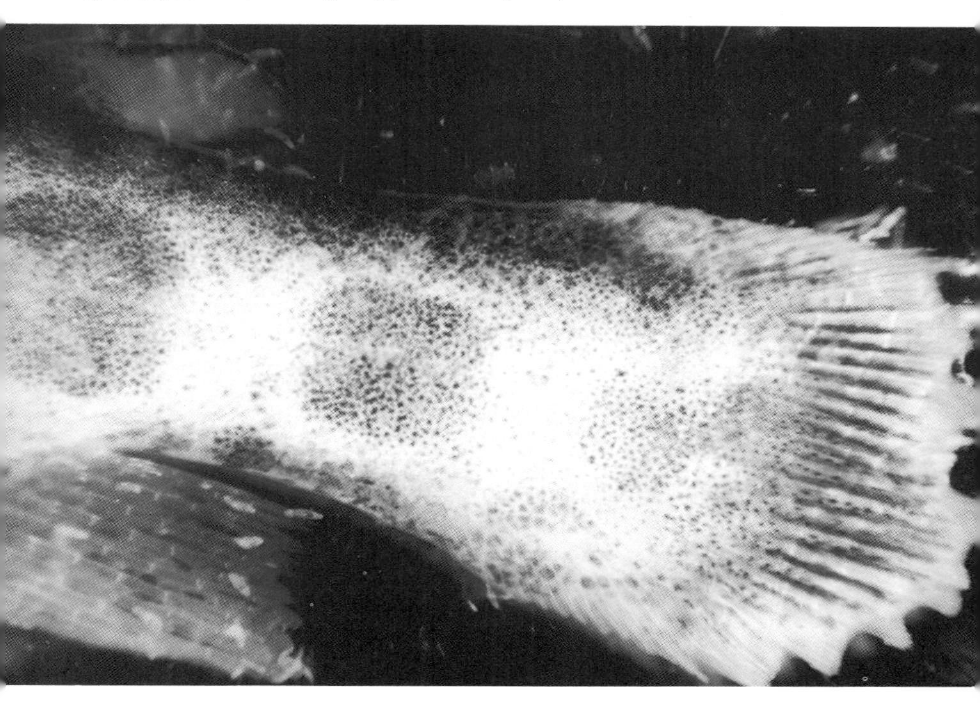

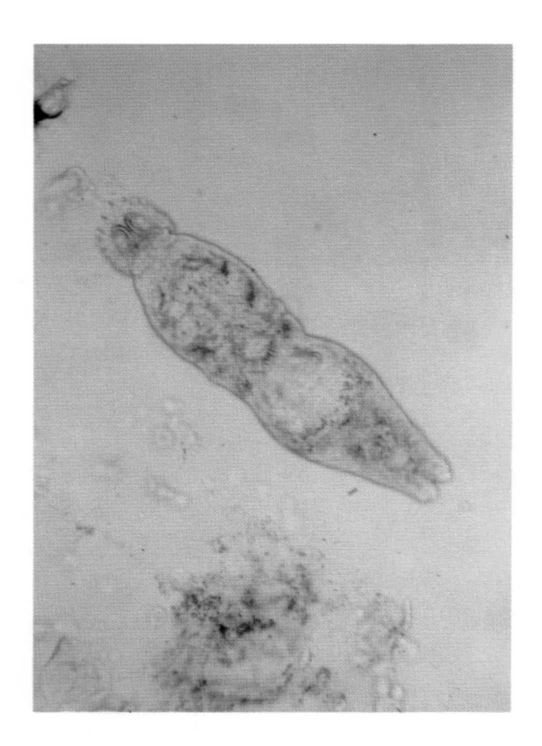

This microphotograph reveals the two principal hooks by which *Gyrodactylus* attaches itself to the host.

Dactylogyrus attached to a gill filament.

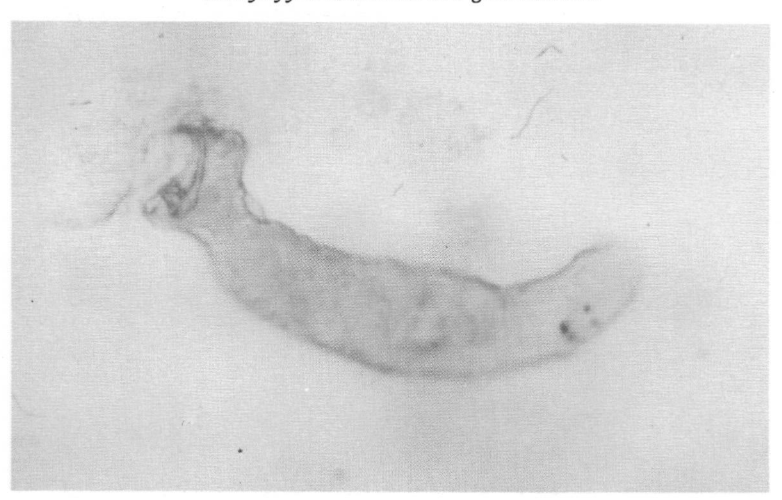

Lernaea, or 'anchor worm'. Length approximately 1 cm.

Argulus. Length approximately 4 mm.

Dactylogyrus

Very similar to *Gyrodactylus*, *Dactylogyrus* can be distinguished by visible eye spots. *Dactylogyrus*, being oviparous, is an egg layer and is commonly referred to as a 'gill fluke', since it attaches itself to the gills in order to feed, gradually destroying the gill filaments. The noticeable signs of an infestation are rapid breathing, and 'flashing' or gasping at the surface of the pond. At this stage, under close examination the gills may appear pale, eroded and perhaps swollen. Healthy gills are blood-red in colour, with well-tapered edges.

Parasitic copepods

Three other parasites that can affect Koi come from a family of crustaceans and are known as parasitic copepods. They are *Lernaea*, *Argulus* and *Ergasilus*. More apparent during the warmer months of the year, they can easily be seen with the naked eye, since the adult parasites are 2–5mm long. The free-swimming larvae stages of these parasites, or even the parents themselves, can be brought to your pond by the introduction of plants, fish, or even water from other sources. Healthy Koi can for a while support a few of any of these parasites, but will eventually be overcome by them.

Lernaea

This parasite is known by Koi enthusiasts as the 'anchor worm'. A fish that appears listless and thin, spending much of its time alone, may be suspected of being infested by *Lernaea*. Sometimes, a Koi can support one or more of these worms temporarily, but will in time become thin and weak as the worm deprives it of nourishment and the site of attack becomes infected with bacteria.

The adult female alone attacks the Koi, securing her anchor-shaped head underneath the skin, usually near the dorsal fin or towards the rear. The body of this parasite can be identified as a thin, transparent worm about 5mm long, sticking out from underneath a scale, the tail end having two egg sacks attached. The eggs will be released into the pond, where they will hatch as free-swimming larvae. In most cases, by the time you have spotted a worm or worms, the eggs will have already been released into the pond and the new larvae developing into adults. After mating, the female seeks a host on which to feed and develop her eggs.

This cycle takes place over a few weeks. Therefore, while treating a Koi with anchor worms, you should take the precaution of eliminating any unattached worms and free-swimming larvae that are in the pond, by using a series of recommended pond treatments.

Argulus

The fish louse, *Argulus,* has much the same effect on Koi as *Lernaea,* making them weak and listless. Koi with these lice can be seen spending a lot of time alone and sometimes swimming erratically as a result of irritation caused by the lice. As with 'anchor worm', only the female attaches herself to the Koi in order to feed and produce her eggs. While feeding, *Argulus* will insert a stylet through the skin and excrete an irritant substance in order to dilute the body fluids. Koi are not only weakened by being deprived of nutrition, but are also affected by the substance used by the lice to enable them to feed.

Argulus appears as a small, 5mm lump of transparent jelly with visible eye spots and attaches itself anywhere on the body or fins. When treating Koi for lice, precautions should be taken to eliminate any lice or larvae which are in the pond.

Ergasilus. Length approximately 2mm. (See p. 106.)

Erosion of the tail caused by bacteria.

Fungus (*Saprolegnia*) growing on injured area.

Symptoms of thermal stress.

Ergasilus

This parasite is known as the 'gill louse' (see p. 103). A Koi infested by it will be seen 'flashing', breathing rapidly or gasping at the surface of the pond. Approximately 2 mm long, *Ergasilus* can be identified on the gills by distinguishable white 'V' shapes, which are the attached egg sacks. The damage caused by this parasite while feeding on the gill filaments will eventually induce respiratory distress. Koi that have become infested should be removed from the pond for treatment, and precautions should be taken to eliminate any lice or larvae present in the pond.

Bacteria

Many types of bacteria can affect Koi. The objective here is not to identify those types, for this involves expensive and sophisticated equipment, but to recognize the various symptoms caused by bacteria. Bacterial infections not only make the Koi weak and listless but can also erode the skin, fins, tail or gills. Blood vessels becoming apparent on fins or tail, reddened areas on the skin, around the mouth or at the tail or fin joints, will reveal a bacterial infection. If the gills become infected, the first signs will be respiratory distress. On examination, the gills could appear swollen, pale and possibly eroded.

The build-up of bacteria that can affect Koi is generally caused by poor water conditions due to the accumulation of organic debris. The infection of an open wound or abrasion caused by a sharp object in or around the pond, the breakdown of the natural defence mechanism of the Koi through stress caused by rough handling, rapid temperature changes or parasite attack, can all leave a Koi vulnerable to bacterial infection. These possibilities should all be considered if an infection is recognized as the cause of illness, and the necessary steps taken to prevent a recurrence of the disease.

Fungal growth

The growth of fungus (*Saprolegnia*) can follow injury, stress, parasite attack or simply poor water conditions. It can also grow on sunken debris and fish waste, so maintenance must be rigorous. Fungus, which destroys the area to which it is attached, appears on Koi as a cotton-wool-like growth on the skin, fins or gills, and is indeed commonly known as 'cotton-wool disease'. If the gills become covered with fungus, the Koi will show signs of respiratory distress.

Sudden drops in water temperature can cause small white lumps to appear on the skin or fins of Koi. This is known as 'carp pox' and is considered harmless. It is not uncommon to find a small white lump (a 'cold spot') on newly imported Koi that have undoubtedly experienced severe drops in water temperature while travelling. This should not be confused with fungal growth, for cold spot associated with carp pox usually remains the same size and will eventually disappear as the water temperature rises.

Swim-bladder disorders

The swim-bladder controls the Koi's ability to rise to the surface, dive to the bottom or remain at any level in the pond. Any disorder in this organ will cause the Koi to struggle to reach the surface only to sink again, to struggle to dive to the bottom only to float to the surface, or to lie or swim at an unusual angle, either at the surface or bottom of the pond.

Swim-bladder disorders can be caused by indigestion or constipation. Feeding should be stopped, and if after a few days there is no apparent improvement, the affected Koi should be removed from the pond for treatment.

Thermal stress

This is a condition rather than a disease, making the Koi thin and listless. Fish that have undergone continual temperature changes can develop thermal stress. Related to this condition is hollow-back disease (dystrophy of the back). This condition is noticeable by the apparent degeneration of the back muscle, which eventually makes the fish appear thin, with a concave back and a large head.

Koi suffering from thermal stress or hollow-back disease can be cured if kept under warmer conditions and fed with food containing antibiotics, to prevent bacterial infection.

Note A fish can also appear thin, not only through the symptoms mentioned above, but also through loss of appetite and/or a lack of oxygen, if there is a gill infection or infestation by protozoan parasites.

Dropsy

This symptom makes a Koi appear bloated with its eyes and scales sticking out rather like a pine cone. There are many causes, but in general, it is due to poor water conditions and at first affects only part of the Koi, before progressing rapidly to the rest of the body. At this advanced stage it will cause respiratory distress and make the Koi swim erratically before dying. Dropsy is difficult to cure, but can be treated at an early stage by a short-term bath, four hours or longer if necessary, containing 25 g of natural salt per 5 litres of water.

Hole disease

Caused by poor water conditions that can favour the appearance of bacteria and the protozoan parasite *Epistylis*, this disease first manifests itself as small, white lumps of about 3–4 mm in diameter. The area around the infection will become red and congested as a hole appears in its centre. As the disease progresses over several weeks, the parasites dissolve away the scales and skin at the site of attack, which is usually on the Koi's side. If the condition is left untreated, the parasites will eventually make a hole in the flesh. At an early stage, short-term baths containing malachite green and natural salt may bring about a cure. In its advanced stage, however, hole disease is difficult to cure. If expensive Koi are affected, it is advisable to ask a good veterinarian or an experienced Koi keeper to administer an intra-peritoneal or intra-muscular injection of a suitable antibiotic.

Symptoms of hole disease.

Preparation for treatment

A diseased Koi will have to be removed from the pond for treatment, and to prevent the disease recurring the pond itself will have to be treated.

In general, treatments are administered by short-term baths and/or long-term pond treatments. Short-term baths are carried out at the side of the pond by placing the affected Koi into a container filled with the correct amount of pond water to which has been added the required amount of medicine. It is helpful to use an airstone and pump to aerate the solution, replacing the oxygen that some chemicals use while being active.

A long-term pond treatment is carried out by introducing into the pond a low therapeutic dose of medicine, which will have an effect over a longer period. The medicines themselves become inactive after a certain delay, either through evaporation or decay.

Short-term baths can be stressful for some Koi, and great care should be taken to administer the correct therapeutic dose of the medicine prescribed for the correct amount of time. An overdose or too long an exposure to some medicines can prove fatal. This caution also applies to long-term pond treatments: too high a dose can not only have adverse effects on the Koi, but can also destroy the beneficial bacteria in the filter bed.

In general, medicines should be kept in a cool, dry place out of the reach of children and pets. If you administer medicines you should wear rubber gloves, since some of those prescribed can harm the skin. Medicines can be stored for only a certain amount of time, and so it is good practice to replace them periodically.

If a Koi needs to be removed from the pond for treatment, or for any other reason, it should be caught with a special Koi net. These are relatively shallow compared to conventional fish nets and have a close-weave mesh. They avoid bending the fish in two and will not remove any scales by catching them under the mesh. It might be necessary partially to drain a large pond in order to reach your Koi.

To examine a fish more closely, place it in a clean plastic bowl or bucket containing pond water. The Koi's protective mucus coating will stick to and be removed by any dry surface, and so if it is absolutely necessary to handle your Koi, wet your hands first.

For cleaning or treating wounds or removing parasites or fungus, if the affected part cannot be reached, place the Koi on a clean damp cloth. The head can be covered with the cloth, which will help to keep it calm, and then the treatment can be carried out.

Medicines

The medicines mentioned can be supplied by your Koi dealer, a good pharmacist or veterinarian.

Note Medicines used at the recommended dose for Koi can be ineffective or harmful to other species of fish.

Malachite green (zinc-free) oxalate

Used to treat fungal growth and, in combination with formalin, against certain parasites.

Topical application: 0.1 per cent solution.
Short-term bath: 2 ppm (see p. 117) for one hour.
Long-term pond treatment: 0.2 ppm

Malachite green comes in powder form, and to obtain the correct ratios needed for treatments by using ordinary letter scales, apply the following method:

1 Measure out 10 g of malachite green and mix with 1 litre of water.
2 Use this solution as follows:

Short-term bath at 2 ppm: 1 ml of the solution per 5 litres of water.
Long-term pond treatment at 0.2 ppm: 20 ml per 1000 litres of pond water.
0.1 per cent solution for topical applications (removing fungus): mix 1 ml with 9 ml of water.

Note Wear rubber gloves when using malachite green and stop filtration for 12 hours during pond treatment.

Great care should be taken when handling Koi.

Formalin

Toxic, do not inhale. Sold commercially as 40 per cent formaldehyde solution. Used against parasite infestations as follows:

Short-term bath (30 minutes): 1 ml per 5 litres of water.
Long-term pond treatment: 15 ml per 1000 litres of pond water.

Note Short-term baths using malachite green and/or formalin should be aerated. During hot weather, formalin, if used as a long-term pond treatment, can induce oxygen depletion. Therefore, provide additional aeration for the pond.

Sodium chloride (natural salt)

Ideally used in recovery tanks at 25 g (two teaspoons) per 5 litres of water (4–24 hours, or longer).

This treatment can be particularly helpful after treating stubborn cases, or baby Koi, for bacteria, parasites or fungus.

Longer treatment can be made by syphoning off a quarter of the tank water every day, cleaning the bottom and replacing the water with clean, fresh water of the same temperature as that of the tank, adding each time 25 g (two teaspoons) of natural salt per 5 litres of water in the tank. Following these procedures will build up and maintain the saline solution. On the fifth day replace half the solution with clean, fresh water and on the sixth day return the Koi to the pond. This is a treatment for many ailments and ideally suited for use by beginners.

Metrifonate

This insecticide is also known as Dipterex 80, Dylox, or Negavon. The recommended dose for Koi can prove harmful to other pond fish, including golden orfe, but not goldfish, shubunkins, comets, or other varieties of *Carassius auratus*.
Long-term pond treatments at 0.25–0.5 ppm or 1–2 g per 4000 litres of pond water.

To remove and control the evolution of parasites, this insecticide can be used if necessary as a long-term pond treatment at regular two-week intervals during the warmer months, with water temperatures around 10°–15°C (50°–59°F). Higher water temperatures, 18°–20°C (64°–68°F) or more, will require weekly treatments.

Note Metrifonate is ineffective against protozoan parasites and will not harm the beneficial bacteria in biological filters, but will reduce zooplankton at 0.5 ppm.

Handle with extreme caution and wear rubber gloves.

Terramycin

This antibiotic is used in treating bacterial infections and can be conveniently administered by adding it to Koi pellet foods. Ratio: 8 g per 1 kg of food, divided if necessary. Thoroughly mix the Terramycin and pellets together in a clean, dry bowl. To make the Terramycin adhere to the pellets, gradually add 20 ml of fresh vegetable oil per 1 kg of pellets. Distribute the same amount as a normal ration during the early-morning feed for seven days.

Alternative method Add one level teaspoon of Terramycin to 0.5 litres of water, adding the required amount of Koi pellets to soak for 10 minutes. Strain and distribute the pellets on a flat tray and leave to dry thoroughly in a warm place away from children or pets.

Distribute the same amount as a normal ration during the early-morning feed on alternate days for two weeks.

Terramycin can also be used in a prolonged bath at 1 g per 50 litres of water for six to eight days (see p. 120).

For many reasons, including pollution, cost and to avoid rendering biological filters inactive, antibiotics are not used as long-term pond treatments.

Note These pellets can be stored only in a cool, dry place for a maximum of five days.

Mercurochrome

This antiseptic is sold under various brand names as a 2 per cent solution. It can be used to disinfect damaged or infected areas on Koi, by application to a clean, dry surface with a cotton bud.

Roccal

Used in short-term disinfectant baths for minor bacterial infections.
Short-term bath: 2 ml per 5 litres of water for one hour with aeration.

Chloramine T

Although it can be a little difficult to arrive at the correct dose, this antiseptic is particularly active against bacterial infections of the gills and skin. Used in short-term baths of one hour, its activity and success depend largely on its concentration being exact in relation to the pH level and hardness of the water used in the short-term bath.

The pH level and general hardness (GH) can be determined by using the simple test kits that can be bought at any garden centre, pet shop or Koi dealer. Once the pH is established, use the chart below to find the correct dose.

pH	soft water	hard water
6	2.5 ppm	7 ppm
6.5	5 ppm	10 ppm
7.0	10 ppm	15 ppm
7.5	18 ppm	18 ppm
8.0	20 ppm	20 ppm

For example, soft water with a pH level of 7.0 used in a short-term bath will require Chloramine T at a ratio of 10 ppm.

Chloramine T comes in powder form, and to obtain the correct ratios needed for treatments by using ordinary letter scales, apply the following method:
Measure out 10 g of Chloramine T and mix with 1 litre of water. (Divide if necessary.)
Use this solution as follows:
Short-term bath at 10 ppm: 5 ml per 5 litres of water.
Short-term bath at 15 ppm: 7.5 ml per 5 litres of water.
Short-term bath at 18 ppm: 9 ml per 5 litres of water.
Short-term bath at 20 ppm: 10 ml per 5 litres of water.

Short-term baths can be carried out beside the pond. This bath was given to remove parasites, the solution containing a mixture of formalin and malachite green. The latter not only has excellent anti-fungal properties, but also renders this treatment particularly effective against parasites.

Short-term baths must be timed very carefully. The maximum exposure time when using formalin and malachite green is 30 minutes.

Water volume in litres

Some treatments are followed up by long-term pond treatments. To achieve this, you will need to know the exact volume of water in your pond in litres. For a square or rectangular pond, multiply (using centimetres) the length by the width by the depth and divide by 1000.

To calculate the volume of water in a circular pond in litres, multiply the radius by the radius by the average depth, and multiply by 3.14. Divide this total by 1000.

Conversion tables

4.56 litres = 1 imperial gallon (3.78 litres = 1 gallon US)
0.56 litre = 1 imperial pint
1 litre = 1000 ml
1 ml = 1 cc or 20 drops from a standard eye-dropper
1 ppm = 1 part per million
1 ppm = 1 mg in 1 l of water
1 m = 39.27 in
1 m = 3.28 ft
1 m = 100 cm
1 ft = 30.48 cm
1 in = 2.54 cm
°Centigrade ÷ 5 × 9 + 32 = °Fahrenheit

Treatments

Success in treating your fish will rely on many factors, the most important of which are:

1 Experience.
2 Not jumping to conclusions. Unless your fish is in obvious distress, observe it over a period of time.
3 Making your diagnosis correctly and in time, then providing appropriate and precise treatment.
4 Preparing all baths and medicines in advance and carefully planning your actions so as not to stress your fish unnecessarily.
5 Leaving a reasonable interval between the same or different treatments. Failure to do so may prove stressful to your fish and undo any good so far achieved.

6 Baby Koi are difficult to treat.
7 As with other animals, the Koi's constitution will directly influence its response to the treatment you provide.

During all treatments the water used should be of the same temperature as that of the pond. Ideally, for short-term baths the pond water itself should be used. All medicated baths should be covered and aerated by using an airstone and pump; do not feed.

Protozoan parasites: *Trichodina, Costia* and *Chilodonella*

Short-term bath (30 minutes): 1 ml formalin per 5 litres of water with malachite green at 2 ppm.
Apply long-term pond treatment: 15 ml formalin per 1000 litres of pond water.

Ichthyophthirius multifiliis (White spot)

Short-term bath (30 minutes): 1 ml formalin per 5 litres of water with malachite green at 2 ppm.
Apply long-term pond treatment: 15 ml formalin per 1000 litres of pond water with malachite green at 0.2 ppm.
Stop filtration for 12 hours.
Repeat long-term pond treatment each week for four consecutive weeks.

Monogenetic Trematodes: *Gyrodactylus and Dactylogyrus*

Short-term bath (30 minutes): 1 ml formalin per 5 litres of water with malachite green at 2 ppm.
Apply long-term pond treatment: Metrifonate at 0.25 ppm or 1 g per 4000 litres of pond water.
Repeat long-term pond treatment every two weeks for four

weeks with pond water at 10°–15°C (50°–59°F) and each week for two weeks if at 18°C (64°F) or more.
Repeat short-term bath if reinfestation should occur during these periods.

Parasitic copepods: *Lernaea* ('Anchor worm')

To find all the attached worms, examine the infested Koi in a plastic bowl or bucket containing pond water. After locating them, place the Koi on a clean, wet cloth and cover the head. Remove the worms very carefully so as not to leave the heads under the fish's skin, which could leave a scar. With a pair of tweezers, remove the worms by pulling gently but firmly upwards and towards the rear of the Koi and apply mercurochrome to the site of attack. If there are several worms it might be necessary to return the Koi to the water momentarily between removals.
Apply long-term pond treatment: Metrifonate at 0.25 ppm or 1 g per 4000 litres of pond water, every two weeks for six weeks with pond water at 10°–15°C (50°–59°F), and every week for three weeks with pond water at 18–20°C (64°–68°F) or more. If necessary, or as a further precaution, feed Koi on pellet food that has been treated with antibiotics as prescribed under Terramycin (see p. 114).

Argulus (Fish lice)

Apply exactly the same treatment as described for *Lernaea* (see above).

Ergasilus (Gill lice)

Short-term bath (30 minutes): 1 ml formalin per 5 litres of water with malachite green at 2 ppm.
Apply long-term pond treatments as prescribed above for *Lernaea* and *Argulus*, including precautionary measures using Terramycin.

Bacterial infections

Short-term bath (1 hour): Roccal at 2ml per 5 litres of water, followed by salt bath for six days.
Short-term bath (1 hour): Chloramine T at required level.
Repeat treatment after 48 hours, if there are no signs of improvement after four days, and feed on Koi pellets containing Terramycin as described. In case of loss of appetite, Terramycin can also be used in a prolonged bath at 1g per 50 litres of water for eight days. The treatment is ideally carried out in a suitable tank, which must be completely covered and away from direct sunlight. The Terramycin should be added gradually over a period of 12 hours, and after 48 hours the solution should be changed as it will decay and become toxic. Always use old water of the same temperature as that in the tank, or prepare a second tank in advance. The Terramycin can now be added directly after mixing in a little water, and the solution changed again after 48 hours. On the sixth day, syphon off half the solution and replace with clean water. Repeat this on the seventh day and return the Koi to the pond on the eighth. With Koi of great value, it is advisable to ask an experienced veterinarian to administer an intra-peritoneal or intra-muscular injection of antibiotics such as Ampicilline or Gentamycine.

Saprolegnia (Fungus growth)

Place the affected Koi on a wet cloth and cover the head. Carefully remove the fungus by using a cotton bud soaked in a 0.1 per cent solution of malachite green. Try to avoid removing scales. Dry the area that has been affected and apply mercurochrome with a clean cotton bud. Return Koi to pond. If the Koi has been seriously affected, after removing the fungus, place the fish in a short-term bath of malachite green at 2 ppm for one hour before returning it to the pond.

Swim-bladder disorders

Stop feeding the Koi for a few days. If there are no signs of improvement, place the Koi into a tank of shallow water of the same temperature as that of the pond. Add 25g (two teaspoons) of salt per 5 litres of water and take the temperature of the water in the tank. Raise this temperature by 2°C (3½°F) by using a thermostatically-controlled aquarium heater.

Keep the temperature steady until the Koi rights itself and swims in a normal manner. Remove the heater from the tank and allow the water temperature to regain that of the pond before returning the fish.

Epistylis (Hole disease)

Remove the affected Koi from the pond, place on a clean, wet cloth and cover the head. Clean the infection as described for fungal growth by using a cotton bud soaked in a 0.1 per cent solution of malachite green. Place the Koi in a short-term bath of malachite green at 2 ppm for one hour and move to a recovery tank containing salt as described under Sodium Chloride (see p. 113). After treatment, return the Koi to the pond and feed them on pellets containing Terramycin. In its advanced stage, hole disease can be difficult to cure, and where Koi of great value are concerned it is advisable to consult an experienced veterinarian, as described in bacterial infections.

Thermal stress and hollow-back disease

Keep Koi under warmer conditions, feeding them for one week every three weeks on pellets containing Terramycin. Both conditions are difficult and take time to cure. Koi that have been successfully treated should be returned to the pond only at warmer times of the year.

Note As a precaution, before treatment, give short-term bath as described under Protozoan parasites.

Injuries (cuts and abrasions)

Remove the Koi from the pond and place on a clean, wet cloth. Cover the head with the cloth and clean the wound with cotton buds. Dry the wound and apply mercurochrome to the damaged area. Return the Koi to the pond. Apply a second treatment the following day.

If the Koi appears to develop a bacterial infection, feed it on pellets containing Terramycin or apply an alternative treatment.

Recovery tanks

If a Koi has become particularly stressed or weakened through injuries or disease, it can be placed in a recovery tank to convalesce. The tank should be situated in a quiet, shaded place and covered completely to keep the Koi as calm as possible. Filtration and aeration should be provided. Clean the bottom of the tank regularly, replacing the water removed with clean, fresh water of the same temperature as that in the tank. Convalescence can extend over several weeks.

Conclusion

About a third of the fish produced commercially and an extremely large percentage of those in their natural environment, die through either natural selection or disease before they become adult. Very few fish die from old age.

Compared to the progress made in human and animal medicine, research into fish disease is a relatively new field and in certain instances still remains a puzzling subject. Specialists and experienced amateurs alike are coming up with new ideas and relatively similar conclusions, so success in treating disease is very much down to experience and a little luck.

Do not be discouraged if you are not successful in treating disease the first time. Information and advice from other Koi keepers should not go unheard and books from writers who perhaps have had the same experience and have also been successful in their research, should not go unread.

Recommended Reading

For those of you who are interested, there are many other excellent books on Koi. The author recommends the following to readers who wish to know more.

To become totally familiar with the many different varieties, consult Dr Herbert R. Axelrod's *Koi Varieties* (TFH Publications, 1988); and perhaps, from a practical point of view, *A Fishkeeper's Guide to Koi* (Salamander, 1986) by Barry James.

Other books available and worthy of study are Glenn Y. Takeshita, *Koi: for home and garden* (TFH Publications, 1969; new edition 1982); *Understanding Koi,* written and published by David E. Hulse and M. I. George, and *The Practical Encyclopedia of Koi* (Salamander, 1989).

There are also Japanese books available dedicated to Koi appreciation containing mainly high-quality colour prints of prize-winning specimens which are worth vast amounts of money. One of these is *Manual to Nishikigoi* by Takeo Kuroki, one of Japan's leading experts on the subject of Koi, and editor of *Rinko: the International Nishikigoi Specialists' Magazine,* which is well worth reading. Both are published by the Shin Nippon Kyoiku Tosho Co. Ltd, Japan.

Index